THE MULTINATIONAL HISTORY
OF STRASBOURG ASTRONOMICAL OBSERVATORY

ASTROPHYSICS AND SPACE SCIENCE LIBRARY

VOLUME 330

THE MULTINATIONAL HISTORY OF STRASBOURG ASTRONOMICAL OBSERVATORY

Edited by

ANDRÉ HECK
Observatoire Astronomique,
Strasbourg, France

Springer

A C.I.P. Catalogue record for this book is available from the Library of Congress.

ISBN 10 1-4020-3643-4 (HB)
ISBN 13 978-1-4020-3643-9 (HB)
ISBN 10 1-4020-3644-2 (e-book)
ISBN 13 978-1-4020-3644-6 (e-book)

Published by Springer,
P.O. Box 17, 3300 AA Dordrecht, The Netherlands.

www.springeronline.com

Printed on acid-free paper

Printed in the Netherlands.

Table of contents

FOREWORD

Strasbourg Observatory is quite an interesting place for historians: several changes of nationality, high-profile scientists having been based there, big projects born (Hipparcos) or installed in its walls (CDS), and so on.

Most of the documents circulating on the history of the observatory and on related matters have however been so far poorly referenced, if at all. Occasionally some elements are definitely wrong, such as even the founder's name in a recently printed leaflet. As it can be expected, such errors are unfortunately carried over in all kinds of publications copying each other. This made necessary the compilation of a volume such as this one, offering fully-documented historical facts and references on the first decades of the Observatory history – until roughly the directorship of Lacroute.

The book starts with an introductory chapter by the Editor, followed by two contributions, by Wolfschmidt and by Duerbeck respectively, dealing extensively with the German periods and reviewing people and instrumentation, while another paper by Duerbeck is more specifically devoted to World War II times. The subsequent chapter by Débarbat describes the development of astronomical instrumentation at Strasbourg Observatory during the French period between World Wars.

Next, three contributions are devoted to specific astronomers: by Duerbeck on Wislicenus, by Duerbeck & Seitter on Wirtz, and by Bacchus on Muller.

Switching to big projects, two chapters by Heck and Jung deal with the genesis and the first years of the Stellar Data Center (CDS) installed at Strasbourg Observatory, while Kovalevsky recalls that the Hipparcos Spacecraft was born there.

Then Seggewiss reminds the links of the Observatory with the Astronomische Gesellschaft, starting with the inauguration of the place. Follow a discussion by Bodenmuller & Heck of the visibility of the Observatory in the regional press around the beginning of the 20th century and two notes by Heck on respectively the Coronelli globe of the Observatory and Danjon & Rougier's first spectroscopic studies of the green flash from Strasbourg.

Several appendices conclude the book: lists of Observatory Council members (Heck & Bruneau), of CDS Council members (Bruneau & Heck) and of Observatory scientific personnel (Traut, Heck & Duerbeck), as well as of the institutional publications (Vonflie & Heck) until Year 2000.

The Editor acknowledges the assistance of all people who helped him putting this book together, in a way or another, and in particular of Mmes & Messrs. P. Abraham, P. Bacchus, M. Barnoud, S. Bengel, W. Bodenmuller, B. Brejon de Lavergnée, Ch. Bruneau, D. Carita, G. Chester, B. Corbin, N. Cramer, S. Débarbat, W. Dick, P. Ditz, Chr. Douyère-Demeulenaere, P. Dubois, H.W. Duerbeck, A. Fresneau, J. Garnier, Fl. Greffe, M. Haller, M. Hamm, R. Hellot, E. Høg, J. Jung, Th. Keller, J. Kovalevsky, J. Krautter, D. Kuute-Puers, Fr. & J.M. Lacroute, S. Langenbacher, J. Levy, L. Ludes, L. Maison, J. Manfroid, J. Marcout, B. Moya, Ch. Muller, F. Ochsenbein, H. Richard, C. Schohn, E. Schweitzer, W. Seggewiss, W.C. Seitter, M. Sessa, I. Tomdus-McLotte, B. Traut, C. Turon, M. & Ph. Véron, Ph. Vonflie, F. Woelfel, G. Wolfschmidt, as well as the archivists of the Académie des Sciences de l'Institut de France.

Finally, it is a very pleasant duty to pay tribute here to the various people at *Kluwer Academic Publishers* who have enthusiastically supported this volume.

The Editor
'La Cruz Verde'
March 2005.

STRASBOURG ASTRONOMICAL OBSERVATORY AND ITS MULTINATIONAL HISTORY

A. HECK
Observatoire Astronomique
11, rue de l'Université
F-67000 Strasbourg, France
heck@astro.u-strasbg.fr

Abstract. Strasbourg Astronomical Observatory changed nationality several times since its foundation in the 19th century. This chapter outlines the observatory history over roughly a century and introduces the directors together with a few high-profile scientists having been based there or associated with the institution during that time. The major instruments are presented as well as several big projects born (Hipparcos) or installed (CDS) in the observatory. Pointers are provided towards other chapters of the book.

1. Introduction

Strasbourg Observatory is quite an interesting place for historians: several changes of nationality, high-profile scientists having been based there, big projects born or installed in its walls, and so on.

Most of the documents circulating on the history of the observatory and on related matters have been so far poorly referenced, if at all. Occasionally some elements are definitely wrong, such as even the founder's name in a recently printed leaflet. As it can be expected, such errors are unfortunately carried over in all kinds of publications copying each other. This made necessary the compilation of a volume such as this one, offering fully-documented historical facts and references.

Sect. 2 of the present chapter recalls a few pecularities of Alsace, that region that switched borders several times in three quarters of a century. If it is French today, life there is not quite the same as in the rest of France – something surprizing in a country where centralization and uniformization

A. Heck (ed.), The Multinational History of Strasbourg Astronomical Observatory, 1–61.

have been the rule for so long. The founding of the observatory is then introduced together with its main features.

Sect. 3 is devoted to a gallery of portraits of the successive directors from August Winnecke to Pierre Lacroute where we are stopping our historical considerations. Subsequent times are too close for a proper review. Their appreciation is left to future generations.

The main historical instruments of the observatory as well as a few noteworthy projects are then reviewed in Sect. 4, while Sect. 5 introduces a few additional scientific personalities associated with the institution. An Epilogue then refers again to the local context.

It should be kept in mind that this chapter does not aim at exhaustivity. It is deliberately centered on what we consider as key features and characters. We encourage readers to peruse the other chapters of this book and to check out the various bibliographical references therein.

2. The Regional Context

Strasbourg, capital of Alsace. A strong regional identity. A eventful recent history with several changes of nationality in three quarters of a century. This got unavoidably to be reflected in the observatory history and in the turnover of its personnel.

2.1. LOCAL PECULIARITIES

One of the last and continuing bestsellers in the region is the book by Laurence Winter (2000) entitled *"Ciel! Mon mari est muté en Alsace ..."* [1]. The subtitle is more explicit on the actual substance of the book: *"Petit manuel de comportement à l'usage des nouveaux arrivants pour leur éviter impairs et déconvenues"* [2]. Full of anecdotes, that volume (Fig. 1) is read with the greatest pleasure, by Alsatians and non-Alsatians alike. Some locals were confessing to feel even more Alsatian after reading the book!

Many French people, and *a fortiori* foreigners, are unaware that the law in Alsace is not quite the same as in the rest of France. A lawyer may plead in Strasbourg, Colmar or Mulhouse only after getting a specific qualification in "local law".

In the same way, locomotive drivers must have received a special training since, in Alsace, trains run on the right like in Germany. Lights and signs are thus positioned on the other side of the tracks compared to their location in *France de l'Intérieur* (Inner France). The tracks themselves pass over each other at the level of the old border.

[1]Heavens! My Husband is Tranferred to Alsace ...
[2]Small Behavior Manual for Newcomers to Save Them Blunders and Disappointment.

Figure 1. The book by Laurence Winter (2000).

Alsace is also preserved ground for the chimney sweepers who recently protested violently against invading colleagues from the nearby Belfort Territory (now a separate French *Département*) who were not holding the necessary degrees nor obeying the same guild rules.

In the continuation of the "Concordat" dating back to the Napoleonic Imperial decrees of 1801-1802, the priests and other cult ministers are paid by the State, which is not the case in the rest of the country. Alsace enjoys also two additional holidays: the Good Friday (Friday preceeding Easter) and the 26 December (day following Christmas), something which is always disturbing national tourists, surprized to find shops and official buildings closed those days.

The payroll of civil servants shows a line corresponding to an 'indemnity for administrative difficulties': a very small amount today as it has not

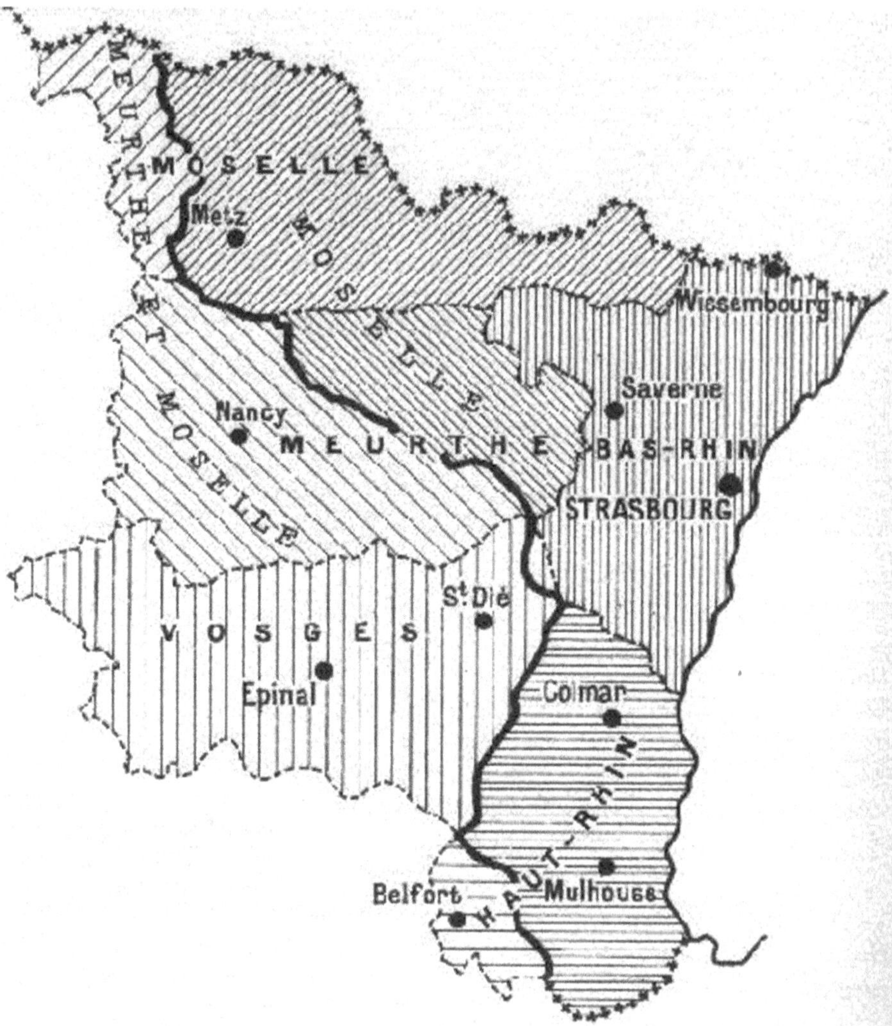

Figure 2. The annexed Alsace-Moselle area, pictured with narrow hachures, East of the new border after the Franco-Prussian war of 1870-1871.

been reevaluated since long. When created, it represented however a non-negligible bonus and a good incentive to motivate 'French' civil servants to come and work in a region where their language was not commonly spoken at that time[3].

[3]In the 1970s, we witnessed people from "Inner France", in particular educated Parisians, repeatedly calling "illiterate" people from Strasbourg (imagine how were then considered people from the countryside) where Alsatian – or French with a heavy accent – was frequently spoken spontaneously in everyday life contact.

Figure 3. Wilhelm I. (Berlin, 1797-1888), the founder of Strasbourg Astronomical Observatory. Crowned King of Prussia in 1861, he was proclaimed German Emperor on 18 January 1871 in the *Galerie des Glaces* of the Versailles Castle before the very end of the Franco-Prussian war.

2.2. A BIT OF HISTORY

At the outcome of the Franco-Prussian war of 1870-1871, France lost Alsace (minus Belfort and a small territory around the city) as well as a part of Lorraine including the city of Metz (Fig. 2). That region, Alsace-Moselle, was improperly named *Elsaß-Lothringen* (Alsace-Lorraine) by the German empire – an incorrect denomination that it is still found today in many books including tourist guides.

As so often in the course of History, the new authorities decided to make a showcase out of the newly acquired region and in particular out of its capital Strasbourg. New spacious and structured quarters were built, roughly East of the old town (Fig. 4). They are still called today the 'Wilhelminian Quarters' from the name of the new masters, the Emperors Wilhelm I. (1797-1888, Fig. 3) and Wilhelm II.[4] (1859-1941) who ruled until the end of World War I (WWI).

[4]Wilhelm II. was in fact the grandson of Wilhelm I. who died in March 1888 and was succeeded by his son Friedrich III. who reigned only a couple of months before dying in June 1888. Wilhelm II. was Friedrich III.'s son.

Figure 4. Excerpt from a 1899 map of Strasbourg showing the Eastern extensions (Wilhelminian quarters). The observatory Big Dome is the cross near the picture center.

Figure 5. Current view of some Wilhelminian university institutes built at about the same time as Strasbourg Observatory. On the left, the imposing outline of Strasbourg Cathedral dominates the old town. The back of the *Universitätspalast* is visible near the center of the picture, at the end of the alley. The observatory is behind the photographer. (Phot. J. Marcout, © Obs. Astron. Strasbourg)

The *"Dictionnaire historique des rues de Strasbourg"* [5] (Moszberger *et al.* 2002) offers interesting details:

> *"Adopted in 1880, the extension plan for Strasbourg, established by Head Architect Jean-Geoffroy Conrath graduated from the* École des Beaux-Arts de Paris *and left at his position by the German administration, increases roughly threefold the town area, from 232 to 618 hectares [...]. Conrath's ambitious design, taking also into account suggestions from August Orth, an architect from Berlin, summonned up during decades all energies and resources. One estimates to one Billion Marks of the time the spending by the State as well as by the City and private owners. The plan is carried out progressively, encounters some difficulties with most streets having to wait ten or twenty years to be completed. But in the end, at the dawn of WWI, the new town has risen from the wasteground."*

The new city extension included what one would call today a university campus with a 'palace' and a whole range of institutes (Fig. 5), together with an astronomical observatory.

[5]Historical Dictionary of the Streets of Strasbourg.

Figure 6. View around 1880 of the *Kaiserliche Universitäts-Sternwarte Straßburg.* A few pathes and young trees of the Botanic Garden are visible in the left foreground. The traces left by cartwheels heading towards the building on the right *(Bâtiment Sud)* mark the future *Universitätsstraße.* (© Obs. Astron. Strasbourg)

The city became French again, as well as the whole of Alsace-Moselle, at the end of of WWI (1918). In 1940, the area was annexed by the Nazi regime at the outbreak of WWII. The French university had already been moved to Clermont-Ferrand. German staff run the activities in Strasbourg during the conflict. The region returned to France at the end of the war (1945).[6]

2.3. THE IMPERIAL OBSERVATORY

Thus the decision to set up an astronomical observatory in Strasbourg just after the French-Prussian war of 1870-71 is to be seen as a political move,

[6]It is generally little known that other neighboring regions, such as the Belgian 'Eastern Cantons', had the same fate during WWII: annexion, enlistment of men in the German armed forces (and sent to the Eastern front), execution of rebels with the families sent to concentration camps, girls having sometimes the option to join military brothels. In the case of Belgium, German remained the regional language after the conflict and the third official language of the country (together with Dutch and French). In France, the presence of several Alsatian elements in the division *Das Reich* moving Northwards towards Normandy in June 1944 and killing several hundreds of civilians (including eleven evacuated Alsatians) in Oradour-sur-Glane (Limousin) has been the source of multiple polemics and heated debates in the country. It is still a delicate matter nowadays.

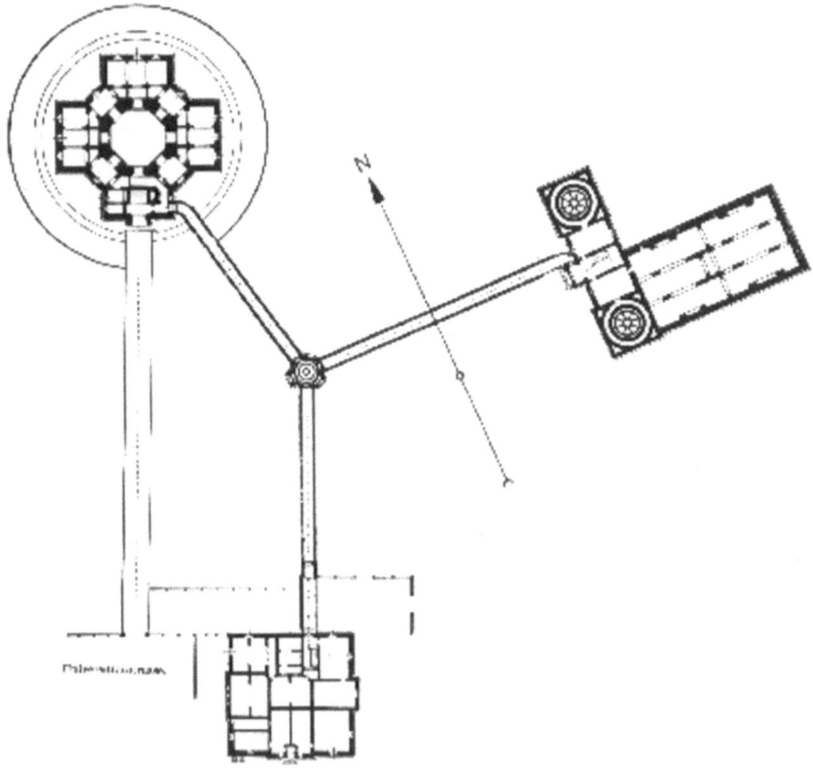

Figure 7. Blueprint of the original buildings of Strasbourg Observatory (Becker 1896a). (© Obs. Astron. Strasbourg)

in line with the construction of an imperial university and of whole new quarters in the city. The construction took place between 1877 and 1880 under the direction of Architect Hermann Eggert. The buildings were inaugurated in September 1881 with a General Assembly of the *Astronomische Gesellschaft* [7] (see *e.g.* Seggewiss 2005).

As shown on Figs. 6 & 7, the observatory consisted from the start of several elements interconnected by covered corridors, still in existence and allowing the personnel to move between the buildings while being protected from the bad weather.

The most emblematic building, the Big Dome, is located at the end of

[7]Founded in 1863 in Heidelberg with an asserted international vocation, the *Astronomische Gesellschaft* (AG) is one of the oldest astronomical societies, second only to the *Royal Astronomical Society* (RAS) founded in 1820 (Pfau 2000). Note that the AG membership includes about 2% amateur astronomers while, in the RAS, they are estimated to make about 55% of the membership (Heck 2000).

Figure 8. Construction of the Eastern Building. (© Obs. Astron. Strasbourg)

Figure 9. Current aerial view of Strasbourg Observatory. The Eastern building, added in 1933 and heightenend in 1958, is visible at the upper right in the continuation of the meridian building. (Phot. J. Marcout, © Obs. Astron. Strasbourg)

Figure 10. Current view of the Big Dome. One of the modern extensions is clearly visible. (Phot. J. Marcout, © Obs. Astron. Strasbourg)

the *Universitätsallee* [8], itself in the continuation of the perspective linking the *Kaiserpalast* [9] designed also by H. Eggert and the *Universitätspalast* [10], conceived by Otto Warth.

The other original elements were (for details, see Becker 1896a as well as Müller 1975, pp. 170-175):
– a residential building (Bâtiment Sud/South Building) for the Director, including offices, at the corner of the inverted-**L**-shaped University Street;
– an observational unit located East of the previous ones and including two smaller domes and two meridian rooms.

The observatory is located next to the Botanical Garden, also part of the urban extension. Anecdotes of the time report how employees were complaining for having to go and work "out of town" – a distance walked in only a few minutes. What would those persons have said of our current campuses, often exiled way out of the cities!

Later on, an office building with workshops in the basement (Bâtiment Est/Eastern Building) was built in 1933 between the meridian rooms and the current *Rue de l'Observatoire* (Observatory Street, Figs. 8 & 9). A

[8]Or *Allée Universitaire* (University Lane), not to be confused with the nearby *Rue de l'Université* (University Street).
[9]Currently *Palais du Rhin* (Rhine Palace).
[10]Today *Palais Universitaire* (University Palace).

TABLE 1. Strasbourg Observatory Directors. The period covered by this chapter ends roughly with the directorship of Lacroute.

1872-1886	A. Winnecke
1882-1886	W. Schur (a.i.)
1886-1887	H. Kobold (a.i.)
1887-1909	E. Becker
1909-1919	J. Bauschinger
1918-1919	A. Baldit (a.i.)
1919-1929	E. Esclangon
1929-1945	A. Danjon ($\Rightarrow$ Clermont-Ferrand)
1941-1944	J. Hellerich (Strasbourg)
1946-1976	P. Lacroute
1976-1987	A. Florsch
1987-1988	D. Egret (a.i.)
1988-1990	A. Heck
1990-1995	M. Crézé
1995-2000	D. Egret
2000-	J.M. Hameury

story was added in 1958. Much more recently, extension modules (offices, auditorium) were added at the "armpits" of the Big Dome cross (Fig. 10). Those additions, albeit approved by the authorities in charge of historical buildings, will most likely not age in the same way as the original building. Therefore they could be removed only by leaving scars at the joints.

Additional internal modifications were brought later on, such as installing a planetarium in one of the meridian rooms in 1982. The second meridian room now hosts the new observatory library inaugurated there in 1995. External adaptations consisted mainly in an entrance module and offices for the planetarium as well as a junction (offices, stairs) between the meridian building and the Eastern building. Of course, we shall not detail here the internal restructurations of offices, irrelevant to our purpose.

3. The Directors

3.1. AN EVOLVING FUNCTION

Table 1 gathers together the successive observatory directors. A couple of comments are in order before commenting various personalities.

First of all, the by-laws regulating French research institutions such as the observatories underwent several changes, especially after the 1968

Figure 11. Strasbourg Observatory shortly after its construction (rare buildings in the nearby streets). (© Obs. Astron. Strasbourg)

events that rocked European universities. Following these, directors in particular were no more nominated in their position for the rest of their active life, but elected for limited terms. This explains the high turnover of directors in the last decades. Lacroute (whose directorship was the longest one, spanning 30 years) went in fact through the post-1968 transition and was reelected until his retirement[11].

Second, and given the way scientific activities are organized and financed nowadays, both nationally and internationally, one must keep in mind that people currently in charge of observatories are much more administrators than directors.

3.2. THE FIRST GERMAN PERIOD

The creation of an observatory implies the nomination of a first director. Thus August Winnecke (1835-1897, Fig. 12) was put in charge of building an observatory with a novel design. According to Architect Eggert (quoted by Mayer 1894), *"For laying out the group of buildings, as well as for arranging each of them in particular, Prof. Winnecke's program was*

[11]Becker's directorship ranks second with 22 years.

Figure 12. August Winnecke (1835-1897), the first Director. (© Astron. Gesellschaft)

determining." Among the innovations, the dissociation of instrument pillars from the building foundations deserves a special mention, as well as the separation of the various areas of activities (lodging, observing, working).

Winnecke was appointed Director for Strasbourg Observatory in 1872. He had been earlier at Pulkovo Observatory that he joined in 1858 and where he became Deputy Director in 1865. In 1864, he had married a niece of Otto Wilhelm Struve[12] who was in charge of Pulkovo since the retirement of his father in 1862. In 1869, Winnecke had been elected Secretary of the *Astronomische Gesellschaft* (AG, see Footnote 7), a position he was going to hold during a dozen years, *i.e.* until September 1881 when Strasbourg Observatory was inaugurated with an AG meeting.

Born in Groß-Heere (Hildesheim), Winnecke had studied astronomy in Göttingen, Berlin and Bonn. He worked with eminent scientists such as Gauß (1777-1855), Encke (1791-1865) and Argelander (1799-1875). He was an excellent observer with ten comets discovered between 1854 and 1877 and numerous stellar observations. In Strasbourg, he initiated an astrometric survey (precise positions) of nebulae which would be completed only in 1911 (by Kobold and Wirtz).

[12]Otto Wilhelm Struve (1819-1905) belongs to a dynasty of astronomers spanning four generations and starting with his father Friedrich Georg Wilhelm von Struve (1793-1864) who set up Pulkovo Observatory (Saint Petersburg).

Figure 13. Strasbourg Observatory on a late 19th-century or early 20th-century postcard. The "i.E." (meaning *im Elsaß*) after the city name removes any doubt on the localization of the place and, in particular, avoids any possible confusion with the other German *Strasburg* (Uckermark), some 140km North of Berlin. Note the surrounding vegetation that could let an unaware eye to believe the location is in an equatorial country.

Winnecke was extremely busy[13] on top of supervising the observatory construction and the instrumental furnishing, not to forget the recruitment of personnel[14]. His family was not spared by tragedy as the eldest son died accidentally in January 1881. Multiple trips and the additional charge of University *Rektor* ended by wearing out Winnecke's health.

As he fell seriously ill in 1882, the observers Wilhelm Schur (1846-1901) and Hermann Kobold (1858-1942) acted as deputies. In 1886, Schur was nominated Professor and Director in Göttingen. Then the Faculty requested the Ministry to provide a new director, a petition that was satisfied with the appointment of Ernst Becker (1843-1912, Fig. 14), born in Emmerich am Rhein, educated in Berlin and Director in Gotha since 1883.

Hugo von Seelinger, Professor and Director in Gotha, then in Munich, had been approached, but he finally declined the offer with the following comment (quoted by Wolfschmidt 2005):

> *"Strasbourg Observatory is so incredibly, so beautifully well set up that I hold it for the best institute, the most adequate in all Germany and in all Austria. It is with deep regret that I have to renounce to take advantage of such a treasure for my scientific investigations."*

[13]For details, see Hartwig (1898).
[14]Wolfschmidt (2005) indicates that no member of the observatory was Alsacian then.

Figure 14. Ernst Becker (1843-1912), the German Director with the longest term. (© Astron. Gesellschaft)

Becker took up his duties as Ordinary Professor of Astronomy and Director of Strasbourg Observatory in December 1887. He requested admission as Emeritus in Spring 1909, essentially because strong pains were diminishing his working abilities. He retired to Freiburg im Breisgau where he died three years later. Becker had been recognized as a talented calculator, but he had also built up a solid observational experience in Leiden, Neuchâtel and Berlin at the beginning of the 1870s.

In his obituary, Jost (1913) describes Becker as being no reformer on his own. Besides the many activities imposed by the directorship (including teaching from which he did not hold much satisfaction), Becker nevertheless initiated the publication of the *Annalen* of the observatory. He also conducted himself many observations of circumpolar stars at the altazimutal instrument (see Sect. 4) after the discovery of the fluctuations of the Pole in 1891. He had also to undertake several campaigns of relative gravity measurements in Alsace-Moselle, a reminiscence that, in the past, Strasbourg Observatory was in charge of both astronomy and geophysics.

Born in Fürth (Mittelfranken) and educated in Munich, Julius Bauschinger (1860-1934, Fig. 16) succeeded Becker. He was then heading the Berlin Astronomisches Rechen-Institut, a position coupled to a chair as Ordinary Professor of Theoretical Astronomy. Stracke (1934) mentions that it was not easy for Bauschinger to leave Berlin, but Strasbourg had

Figure 15. The Big Dome and the Botanic Garden before World War I (cf. Fig. 53). The fenced corner at the right border of the picture corresponds to the current *Rue Goethe*. (© Obs. Astron. Strasbourg)

some serious strong points: the observational equipment remarkable for the time; the proximity of the Black Forest and the Vosges; the easier life in a mid-size town; and finally the appreciated directorial residence surrounded by gardens offering larger moving space to the Director and his family. Bauschinger took up duties on 1st April 1909.

New programs were undertaken, such as the determination of positions for nebulae at the Large Refractor (see Sect. 4) as well as measurements of double stars at the meridian circle. Geodetic observations were continued in Alsace-Moselle at the request of the government.

The first World War (WWI) restricted and disturbed the activities of Bauschinger[15] and of his collaborators: the university was used as a military hospital; troops were camping in the observatory gardens and in the Big Dome; it was even officially planned to dig mass graves in those gardens in case of siege! When, in January 1919, Bauschinger had to leave Strasbourg with the permission to take away only his own personal observations, he went to Munich. In 1920, he was called to Leipzig (succeeding H. Bruns) where he stayed until 1930.

[15]Bauschinger also lost a son in 1916 on the front in Northern France.

Figure 16. Julius Bauschinger (1860-1934), Director until World War I. (© Astron. Gesellschaft)

See also Duerbeck (2005a) and Wolfschmidt (2005) on the first German period.

3.3. BETWEEN THE WORLD WARS

The two French Directors between the two World Wars, Ernest Esclangon (1876-1954, Fig. 18) and André Danjon (1890-1967, Fig. 20) had *a priori* parallel careers, but their personalities and scientific profiles were very different. Both were Directors in Strasbourg before being in charge of Paris Observatory. History has retained mainly their Parisian years: Esclangon is remembered as being the father of the *talking clock*, accessible by telephone, that entered public service on 14 February 1933; Danjon has left a formidable mark of authority on French astronomy, on top of his various instrumental developments such as his impersonal astrolabe (described in *e.g.* Danjon 1958 & 1960a).

Esclangon, an astronomer at Bordeaux Observatory and professor at Bordeaux University, was nominated Director of Strasbourg Observatory and Professor of Astronomy at the Faculty of Sciences in January 1919. In his document entitled *Titres et Travaux Scientifiques* (1930), Esclangon explains:

"After the armistice [11 November 1918], I was sent to Strasbourg Ob-

Figure 17. Unusual view of Strasbourg Observatory on another late 19th-century or early 20th postcard. Was the photographer on top of a tree, using a crane or balloon-borne as it was fashionable at the time for taking aerial pictures? In the far are visible the twin spires of the protestant church Saint Paul, known also then as protestant garrison church. Barracks were adjacent to the university campus (cf. Fig. 4), behind the houses on the left of the photograph, together with extensive exercising ground still called *Esplanade* and occupied today by modern housing and new University buildings.

servatory as Director and Professor of Astronomy at the Faculty of Sciences. Reorganizing the observatory in terms of equipment and personnel was then quite difficult because, on one hand, of the industrial and economic disarray, but also because of the extreme difficulties in staffing. Currently that reorganization has been achieved. All services have been restructured; the scientific equipment has been renewed and augmented. Astrophysics holds an exceptional share in the investigations. As to astrometry, far from being abandoned, its means have been improved and tuned to the progress of modern mechanics."

Along the general conditions of the armistice regarding Alsace-Moselle, Strasbourg Observatory had been transferred to France on 22 November 1918. During the war, the personnel had been reduced to a strict minimum, in particular to men whose age or physical conditions had kept away from military activities on the front.

Esclangon (1926) recalls such facts in the opening lines of his review of the new organization of the observatory (including also geophysical aspects): *"Similar circumstances took place in all belligerent countries, dam-*

Figure 18. Ernest Esclangon (1876-1954), the first French Director, in a drawing dated 13 January 1930, *i.e.* just after his arrival at Paris Observatory. (© Acad. Sciences Inst. France)

aging considerably scientific activities that had no immediate applications to warfare." At the end of November 1918, there was only one assistant, a concierge and a mechanics left in the observatory, the Director [Bauschinger] having been obliged to leave a few days after the arrival of French troops in Alsace. Albert Baldit, French officer and meteorologist had been put in charge *ad interim* of the observatory. The German astronomer (Karl Schiller) was "requested" by the new French authorities to stay a few more months at his position in order to continue observing until French staff be hired. The mechanics was also maintained to ensure good maintenance of the instruments. One cannot avoid being struck by the neutrality of Esclangon's discourse and by his consideration for his German colleagues.

"Esclangon was quite an affable man, and he did not speak ill of people. Danjon had no such scrupules.", tells us Jacques Levy (2003). This is very gracefully said. In a letter to André Couder dated 22 September 1930, Danjon releases *"Esclangon is a scoundrel."* in the context of their disagreement on the future location of Haute Provence Observatory (quoted by Véron 2001). Danjon had then just been nominated Director of Strasbourg Observatory, an application on which Esclangon had produced a positive report (Esclangon 1929), in the same way he would do it later on for his succession at the directorship of Paris Observatory (Esclangon 1945).

Figure 19. General view of the observatory surroundings in the twenties (Esclangon 1926). The Rhine (flowing from right to left) is visible in the upper part of the picture and Germany is beyond it. Today the area is totally urbanized, but the green spaces around the Big Dome (center right) have been largely preserved. (© Obs. Astron. Strasbourg)

Apparently Danjon had difficulties to accept divergences with his own ideas, at the professional level as well as on human grounds. When we arrived in Strasbourg, several independent sources were mentioning problems in the career of Alsacian astronomers whose conduct Danjon would have disapproved during WWII[16]. As to the relationship with Esclangon, Levy

[16]With all the usual caution in historical interpretation, it seemed that Danjon did not accept members of his staff, drafted in the French army at the war outbreak and taken as prisonners, remain in Alsace after being offered release by the occupying military authorities – a policy though that is routinely applied by victorious armies as it could be seen again with the recent (2003) invasion of Irak. Fehrenbach (1990) offers however some additional clues on the whole issue. From his comments on p. 239 of his book, it appears that Joseph Huss was the only astronomer who stayed in Strasbourg. Huss always said that, once freed by the Germans, he wanted to remain close to his wife and kids in Alsace. A few pages earlier (235), Fehrenbach (1990) says that – no name specified – an astronomer who had stayed in Alsace came with a German delegation to see Danjon in Clermont-Ferrand and claimed some material Danjon had taken away from Strasbourg Observatory. From the above, one unequivocally concludes that Huss was that astronomer and one can easily understand Danjon surely did not appreciate such a step on which Huss himself might have had little choice. For the rest of his life, Huss repeatedly complained that Danjon had systematically blocked his career after WWII. At this point, it is fair to add that – as far as we could check it – all German instruments

Figure 20. André Danjon (1890-1967), the second French Director. (© Acad. Sciences Inst. France)

(2003) adds:

> *"About Esclangon, [Danjon] was not soft. [Esclangon] had requested Danjon, upon his arrival in Strasbourg, to establish renovation plans for the observatory, plans that were followed and for which Esclangon would have claimed credit (something I do not believe true)."*

Called by Esclangon to Strasbourg as *Aide-Astronome* at the end of WWI, Danjon has been extremely active by carrying out all kinds of observations with the Large Refractor, but also by designing and building new instruments, such as his "cat-eye" photometer or his micrometer based on double-imaging through birefringence (developed with Muller). In 1929, he took part, with Lallemand and Rougier[17], to an expedition for observing the total solar eclipse of 9 May 1929 in Indochina (see Sect. 4).

During his time as Director in Strasbourg, Danjon produced what has been a bible for many: the book *"Lunettes et Télescopes"* (1935) that he wrote with Couder. That period saw also his novel elaborations of passage

had been left in place some twenty years earlier at the end of WWI. See also Winter's book (2000) mentioned earlier (Sect. 2) for a couple of anecdotes on the efficiency of German bureaucracy and archives, police and military officers showing up here and there during WWII with documentary evidence going back to WWI and earlier.

[17]See more on Muller, Lallemand and Rougier in Sect. 5.

Figure 21. Johannes Hellerich (1888-1963), in charge of the German observatory during World War II. (© Astron. Gesellschaft)

instruments and astrolabes that Kovalevsky (1967) qualified as *"greatest advances in positional astronomy"*. *"All in all, he was a really great fellow!"* concludes Levy (2003).

See also Débarbat (2005) on this period.

3.4. WWII AND THE POST-WAR PERIOD

The nearing of WWII induced the transfer of Strasbourg University and of its personnel to Clermont-Ferrand in 1939, something that was actively managed by Danjon who had become Dean in 1935. German authorities were however going to repopulate the Alsacian institution. Thus, on 28 August 1941, Johannes Hellerich (1888-1963, Fig. 21) was nominated Professor of Astronomy and put in charge of the observatory.

Hellerich was working at Hamburg-Bergedorg Observatory when he was drafted at the beginning of the hostilities as ensign in a maritime transportation company based at Wilhelmshaven shipyards. When in Strasbourg, Hellerich ensured several courses (also in Freiburg im Breisgau) and attempted to make best usage of the observational equipment. At the end of WWII, he was interned at Saint-Sulpice-sur-Tarn. He was authorized to go back to Hamburg in February 1946 (Strassl 1963).

Figure 22. Pierre Lacroute (1906-1993), the Director with the longest term. (© Fr. Lacroute)

To the question that many certainly have, one can answer that Hellerich became member of the national-socialist party in 1937, probably to ease his career, but that he obviously was not a fanatic follower (Duerbeck 2005b).

Danjon being called to Paris at the end of WWII, Pierre Lacroute (1906-1993, Fig. 22), an astronomer since 1935 in Toulouse, was nominated Professor in Strasbourg cumulating the function of Observatory Director.

In spite of being a spectroscopist by training, Lacroute decided to go on with the astrometric tradition of the observatory and he modernized the meridian circle (chronograph, photography of the circles, etc.). He also directed the first university computing center and was Dean of Strasbourg Faculty of Sciences from 1949 to 1952.

Lacroute quickly understood the importance of artificial satellites and the gain in precision that could be obtained from astrometric measurements collected outside the terrestrial atmosphere. His idea for an astrometric spacecraft took shape gradually and the Hipparcos satellite was launched in 1989. See more on this in Sect. 4.

It is also during Lacroute's directorship that CDS was installed at Strasbourg Observatory. Initially called Centre de Données Stellaires (Stellar Data Center), CDS would be subsequently known under its current and generalized name as Centre de Données astronomiques de Strasbourg (Strasbourg astronomical Data Center – see Sect. 4).

Figure 23. Jean Jung (1944-), the first Director of the *Centre de Données Stellaires (CDS).* (© J. Jung)

3.5. THE FIRST CDS DIRECTORS

Officially created in 1972[18], the new unit was located in Strasbourg within the scope of a policy of regionalization and of "revitalization" of French provincial observatories.

Jean Jung (1944- , Fig. 23), with whom we were then working at Paris Observatory, had the difficult task to launch quite an innovating project for the time, the utility of which was even questioned by many Fremch astronomers. Fortunately, the structure set up by the Institut National d'Astronomie et de Géophysique (INAG[19]) involved participation from foreign astronomical institutions who were thus *de facto* backing the initiative.

In 1974, Jung decided to redirect his activities and he left astronomy. He was replaced by Carlos Jaschek (1926-1995, Fig. 24), from La Plata Observatory in Argentina, who was then on leave at Geneva Observatory in Switzerland. The official transfer of power took place in 1975. Jaschek was not coming alone. His wife Mercedes was an accomplished spectroscopist and was then scientifically strengthening the very reduced CDS staff.

[18]See Heck (2005) for the pre-CDS years.

[19]Created in 1967, INAG has been renamed in 1985 as Institut National des Sciences de l'Univers (INSU).

Figure 24. Carlos Jaschek (1926-1995), the second CDS Director. During his term, CDS was renamed *Centre de Données astronomiques de Strasbourg* to encompass non-stellar data (solar system excluded). (© A. Heck)

The fifteen years of Jaschek's term saw spectacular progress for CDS (especially at the level of its international notoriety and adoption by projects, institutions, and agencies worldwide) thanks to the reinforcement of its staff, in particular by a member coming from the space research community with extensive European and US connections. CDS then became quickly recognized as an international excellence center and as the only viable unit compared to some concurrent initiatives launched in the first half of the 1980s. Its scope was enlarged to data on non-stellar objects (solar system excluded). Last, but not least, the dramatic evolution in computing and the popularization of electronic networks (Heck 2002) completed the world penetration of the Center.

The current CDS relative abundance of personnel and finances (contracts, etc.) could lead people to easily forget (or to ignore for those who were not involved) the modesty of the initial means (human as well as material ones), the precarity of a status occasionally questioned, as well as the difficulties to get the national decision takers to understand the relevance of the project and the fleuron it was going to be[20]. The clearsightedness, the appropriateness of decisions ensuring the future, and determination of

[20]Such an excellence was first recognized internationally. In France, it really took place with the Hipparcos project (see Sect. 4).

Figure 25. Current view of the Big Dome. (Phot. J. Marcout, © Obs. Astron. Strasbourg)

the initial CDS managers and of their few collaborators are thus even more meritorious.

4. About Instruments and Big Projects

Many institutes and university departments of astronomy/astrophysics can be today extremely active observationally-wise while not operating nor hosting their own instruments. Specific bodies manage multiple ground-based or space-borne facilities welcoming visiting astronomers and/or carrying out observational programmes selected by expert committees[21].

[21]The observational sociology is currently undergoing significant changes: observations are increasingly carried out by resident astronomers optimizing the available time (see *e.g.* Robson 2001 or Comerón 2004, as well as the references quoted therein).

Figure 26. Cover of the inventory book for Strasbourg Imperial Observatory, initiated in 1886. (scan by the author)

When Strasbourg Observatory was founded, the very term of *observatory* was synonym of observations carried out within its walls. Thus the decision to create a new *observatory* implied automatically that it was furnished with instruments from the start.

4.1. A BILINGUAL INVENTORY

From the Franco-Prussian war to the French-German reconciliation, the official discourses in the two countries often consisted of words of hatred and revenge. In such an atmosphere, what could be expected from a new power taking over scientific institutions? To make a clean sweep of the past? To destroy the archives of the previous occupant? Or at least to relegate them to obscure attics and to disregard them forever?

The inventory of the *Kaiserliche Universitäts-Sternwarte Straßburg*, initiated in 1886 (cover reproduced in Fig. 26) is still in our hands. It reveals a moving continuation of entries until the end of the thirties, switching from German to French at the end of WWI.

Thus Fig. 27 shows the inventory table of contents where the German headings are followed by their French translations, sometimes not quite identical. For instance, Fig. 28 illustrates, with a first writing dated 1919, the transition from German to French for entries under Section '**K**' entitled *Schmiede, Schlosser u. Mechaniker-Werkzeuge* (Smithy, Fitter and

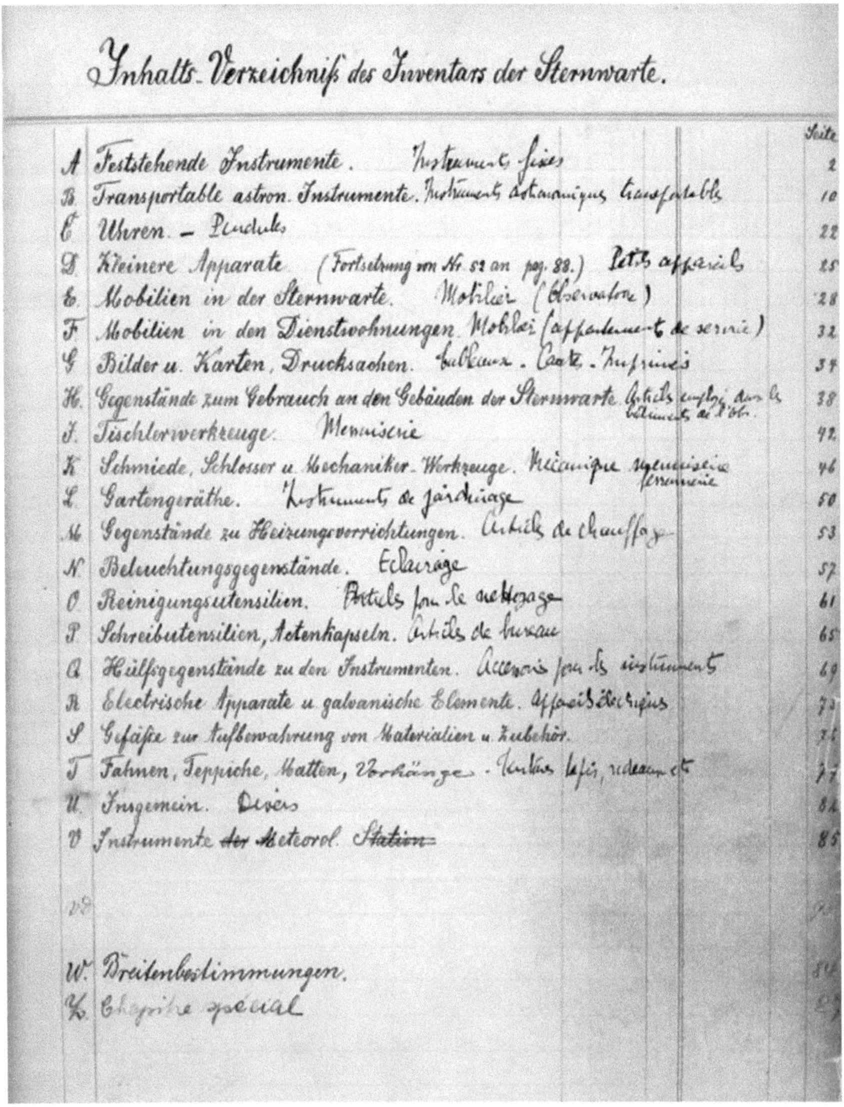

Inhalts-Verzeichniß des Inventars der Sternwarte.

		Seite
A	Feststehende Instrumente. Instruments fixes	2
B	Transportable astron. Instrumente. Instruments astronomiques transportables	10
C	Uhren. – Pendules	22
D	Kleinere Apparate (Fortsetzung von Nr. 52 an pag. 88.) Petits appareils	25
E	Mobilien in der Sternwarte. Mobilier (Observatoire)	28
F	Mobilien in den Dienstwohnungen. Mobilier (appartements de service)	32
G	Bilder u. Karten, Drucksachen. Tableaux. Cartes. Imprimés	34
H	Gegenstände zum Gebrauch an den Gebäuden der Sternwarte. Articles employés dans les bâtiments de l'Obs.	38
J	Tischlerwerkzeuge. Menuiserie	42
K	Schmiede, Schlosser u. Mechaniker-Werkzeuge. Mécanique menuiserie serrurerie	46
L	Gartengeräthe. Instruments de jardinage	50
M	Gegenstände zu Heizungsvorrichtungen. Articles de chauffage	53
N	Beleuchtungsgegenstände. Eclairage	57
O	Reinigungsutensilien. Articles pour le nettoyage	61
P	Schreibutensilien, Aktenkapseln. Articles de bureau	65
Q	Hülfsgegenstände zu den Instrumenten. Accessoires pour les instruments	69
R	Electrische Apparate u. galvanische Elemente. Appareils électriques	73
S	Gefäße zur Aufbewahrung von Materialien u. Zubehör.	[illegible]
T	Fahnen, Teppiche, Matten, Vorhänge. Tentures, tapis, rideaux etc	77
U	Insgemein. Divers	82
V	Instrumente ~~der~~ Meteorol. ~~Station~~	85
W.	Breitenbestimmungen.	86
X.	Chapitre spécial	[illegible]

Figure 27. Table of contents of the inventory. French translations have been added to the original headings in German. (scan by the author)

Mechanics Tools) translated as *Mécanique Menuiserie Serrurerie* (Mechanics Woodwork Locksmithing).

The inventory is obviously not complete for the French years, but entries have been logged until the end of the thirties. This could be put in line with the fact that, at the return of Alsace-Moselle to France, people quite naturally continued using German material installed in the area.

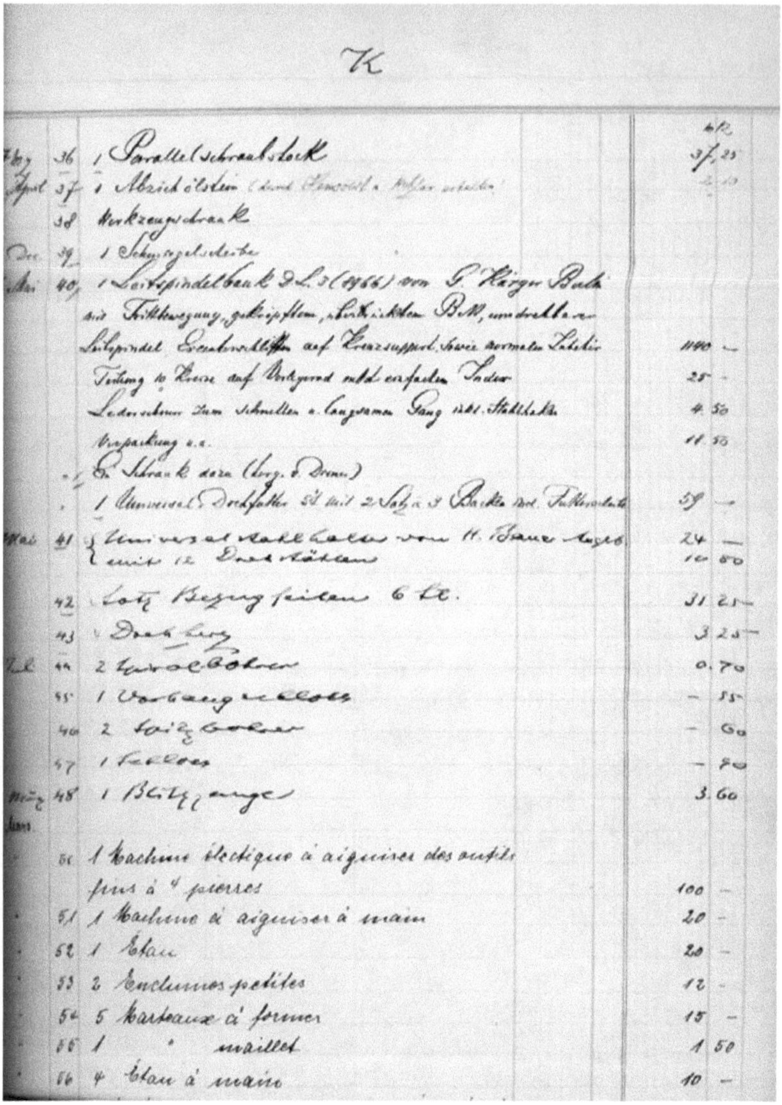
51 1 Machine à aiguiser à main 20 –
52 1 Étau 20 –
53 2 Enclumes petites 12 –
54 5 Marteaux à former 15 –
55 1 " maillet 1 50
56 4 Étau à main 10 –

Figure 28. One of the bilingual pages of the inventory, illustrating the continuity of usage and the switch from German to French after World War I. (scan by the author)

Thus trains continue to this day to run on the right-side tracks as mentioned in Sect. 2. Émile Schweitzer, kingpin of the *Association Française d'Observateurs d'Étoiles Variables* (AFOEV – French Association of Variable Stars Observers), reminded us this was also the case for typewriters and – no joke at all – that the administrations of Alsace-Moselle were then employing *accent clerks* in charge of adding manually French accents on the texts typed on the German machines deprived of such features!

Figure 29. View of Strasbourg Observatory towards the end of the 19th century (Becker 1896a). (© Obs. Astron. Strasbourg)

4.2. OBSERVING INSTRUMENTATION

The first volume of the *"Annalen der Kaiserlichen Universitäts-Sternwarte in Strassburg"* [22] was published in 1896, roughly a quarter of a century after the nomination of the first Director Winnecke and about fifteen years after the inauguration. Becker, the second German Director, describes in that publication not only the buildings, but also the instruments they were housing as well as the observations at the meridian circle for the period 1882-1886. Fig. 29 is reproduced from that volume.

Becker (1896a) offers a good description of the initial instrumentation that appears actually also in the *Inventar* introduced earlier. As emphasized by Wolfschmidt (2005), *"the German instrument-makers were the best on the market at that time"* and they quite naturally equipped the new observatory.

A 132mm Cauchoix passage instrument, built in 1826, was however recovered from a building located *Rue de l'Académie* where earlier astronomical activities had been taking place. A new eyepiece was made by Repsold from Hamburg. The instrument was put in the West meridian room, today occupied by the planetarium. The instrument is currently exposed in the planetarium exhibit area, called the *Crypte aux Étoiles* (Stars Crypt) and located in the basement of the building (Fig. 30).

[22]Annals of Strasbourg Imperial Observatory. See Vonflie & Heck (2005) for a list of institutional publications.

Figure 30. Cauchoix passage instrument built in 1826 which was located the Western meridian room. (© A. Heck)

A 76mm heliometer by Utzschneider & Fraunhofer (Fig. 31) was moreover acquired in 1877 from the Ducal Observatory in Gotha. The following year, Repsold equipped it with a new graduated circle. In 1874, that instrument was part of an expedition to the Kerguelen Islands for the transit of Venus. It was sent to Bahia Blanca (Argentina) for the 1882 transit together with Hartwig (see Sect. 3) and Wislicenus (see Sect. 5) as astronomers from Strasbourg (see *e.g.* Duerbeck 2004). Today that venerable traveller is collecting dust in an observatory storage room.

A 160mm meridian instrument was purchased from Repsold in 1876 and assigned in 1880 to the East meridian room (Fig. 32). It did not move, surrounded today by the new library inaugurated there in 1995 (Fig. 33). That astrometric instrument was extensively used all along the years until the end of the 1960s. Esclangon (1926) indicates that, at the end of WWI, that meridian circle was the only instrument still operational[23]. Observations were carried out, even after Esclangon's arrival as first French Director

[23]The objective set of the Large Refractor (see below) had been stored in the basement to preserve it from potential air bombings.

Figure 31. The 76mm heliometer, a venerable traveller that was part of expeditions (Kerguelen and Argentina) to observe transits of Venus in the second half of the 19th century. (© A. Heck)

(January 1919), by the German astronomer Karl Schiller "requested" to stay to ensure continuity. The first French astronomers (other than the Director) arrived in May 1919 and Schiller left for Germany in August 1919. The German mechanics (Libertus), retained for maintenance, left in June.

The Large Refractor, a 487mm telescope, was built in 1877 by Merz (Munich) who had succeeded Fraunhofer. The mounting itself was manufactured by Repsold in 1880. The instrument was then the largest in Germany (Figs. 34 & 35). In his obituary for Winnecke, Hartwig (1898) enthusiastically commented that:

> *"Whatever be the improvements invented by the quick progress of technology, [the Large Refractor] will always remain a exemplary equipment and will retain its efficiency."*

The instrument is still in the Big Dome, the roof of which was renovated in 1995. It is routinely open to the public and occasionally hosts cultural and musical evenings which – no doubt about that – would certainly be a big surprize to the designers and to the past august users of the instrument ...

Figure 32. Repsold meridian instrument (Esclangon 1926). (© Obs. Astron. Strasbourg)

As visible for instance on Figs. 29 & 36, the building hosting the meridian rooms is topped by two smaller domes. The Northern dome was equipped with a 136mm altazimutal refractor built by Merz & Repsold in 1879 and improved by Bamberg (Berlin) in 1891 (Fig. 37). That instrument is currently dismounted and its elements have been stored away in a corner of the observatory. The Northern dome currently hosts a 600mm reflecting telescope carrying piggyback a *Fliegerkamera* used for observing orbiting spacecraft[24].

As to the Southern dome, it was hosting a 162mm refracting telescope manufactured in 1876 by Reinfelder & Hertel (Munich) and equipped in 1895 by a wire micrometer manufactured by Repsold (Fig. 38). Since the 1980s, that telescope is exhibited at the *Musée de l'Instrumentation Optique* (Museum of Optical Instrumentation[25]) in Biesheim (Haut-Rhin). The Southern dome currently houses a 210mm refracting telescope built by Mailhat. Its objective lenses had been re-shaped in 1952 by Couder & Texereau (Lacroute 1956).

[24]K37 cameras, positioned on the roof of the East building, were also used for that purpose, particularly under the direction of Muller (see Sect. 5).

[25]`http://www.astrosurg.com/euroastronomie/F-68600.htm`

Figure 33. The meridian instrument today, at the center of the new observatory library. (© A. Heck)

The *Inventar* lists numerous other instruments among which a famous 162mm comet seeker built by Merz in 1876 with an altazimuthal mounting set on a mobile chair (Fig. 39). Other comet seekers, small refractors and various instruments were part of the sizable equipment of Strasbourg Observatory in those initial times. They went through the conflicts with uneven fortune. At the end of WWI, various restorations and improvements were brought by Esclangon and his collaborators (Esclangon 1926), the detail of which is of course outside the scope of this overview.

As to the astrophysical equipment largely inexistent at that time (to the exception of an astrophotometer from Gotha Observatory), it is essentially under Bauschinger that Strasbourg Observatory acquired spectroscopic, photographic and photometric devices. Photometers of all kinds multiplied with the French team after 1919.

As to time measurement, critical for activities centered on astrometry, Wolfschmidt (2005) lists:

– a Petit pendulum from the building located *Rue de l'Académie*;

Figure 34. The Large Refractor in the twenties (Esclangon 1926). The movable stairs visible on the right are still existing. The central part can be raised and lowered to position at best the observer. (© Obs. Astron. Strasbourg)

Figure 35. Current ocular end of the Large Refractor. (Phot. J. Marcout, © Obs. Astron. Strasbourg)

– main clocks manufactured by Hohwü & Knoblich (1886);
– a Riefler precision pendulum (1907) who was then the first manufacturer.
There was also a Thomas computing machine manufactured by Burckhardt (1892).

4.3. THE TOTAL SOLAR ECLIPSE OF 1929

That total solar eclipse, visible in Indochina on 9 May 1929, was an opportunity for the French Bureau des Longitudes to set up a national expedition coordinated by General Ferrié[26]. Strasbourg Observatory took part via Danjon (head of group) and Rougier, with the later addition of Lallemand. Such an *extra-muros* observing endeavor was far from being common at a time when intercontinental trips where only possible by boat.

The Strasbourg scientific program was organized around four themes:
– the study of the *Einstein effect* following a method developed by Esclangon (then Observatory Director);
– the determination of contacts;

[26]As an radio pioneer, Gustave Ferrié (1869-1932) developed military applications of the *télégraphie sans fil (TSF)*.

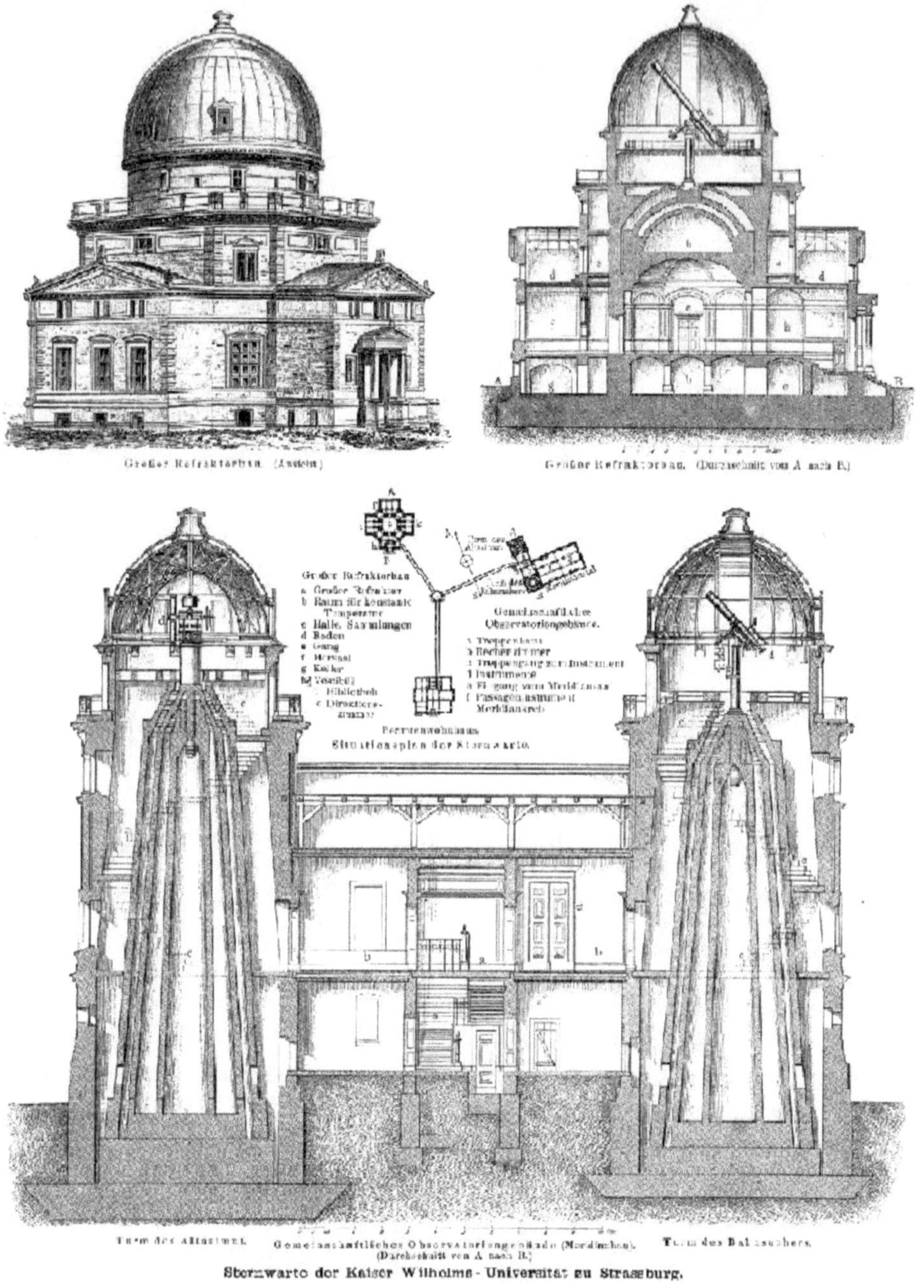

Figure 36. Sketch of Strasbourg Observatory reproduced from F. Küstner's article in the renowned *Meyers Konversations-Lexikon* (1888). A Russian version of that drawing was recently on sale on the Internet.

Figure 37. The altazimutal refractor of the Northern dome (Esclangon 1926). (© Obs. Astron. Strasbourg)

Figure 38. The 162mm refractor of the Southern dome, today at the Museum of Optical Instrumentation in Biesheim. It is pictured here at a intermediary position in a dome located in the observatory gardens. (Phot. J. Marcout, © Obs. Astron. Strasbourg)

Figure 39. Gilbert Rougier using the comet seeker movable along the Big Dome (Esclangon 1926). (© Obs. Astron. Strasbourg)

Figure 40. Strasbourg Observatory station at Poulo Condore for the total solar eclipse of 9 May 1929 (Danjon *et al.* 1938). (© Obs. Astron. Strasbourg)

– the photometry of the solar corona; and
– the study of that corona in the red and the infrared.

The team used a 240mm double (visual/photographic) equatorial refractor lent by Paris Observatory[27] and equipped with photometers built in Strasbourg. The observing station (Fig. 40) on the *Grande Condore* (Poulo Condore archipelago[28]), was located on a sand dune in the middle of a cultivated plain and not far from a ... penitentiary.

From the mission report (Danjon *et al.* 1938), it appears that the bad weather (monsoon reversal) disturbed the initial plans: no first contact; no observable *Einstein effect*, the stellar images being awful and simply unusable. For the rest, in spite of a veil of clouds, the Strasbourg team assisted by navy personnel could take a few good images of the corona in a phase of solar activity maximum.

The infrared photographs (Lallemand) were the first ones of the their kind. The global photometric study of the corona (Rougier) was impaired

[27]The objectives lenses were re-shaped by Couder.

[28]At the Southern tip of the current Viet-Nâm, Poulo Condore archipelago is formed by eighteen islands for a total area of 17km^2. Its history cannot be dissociated from its *bagne* (convict prison) located on the biggest island, the *Grande Condore*. See for instance the book by Demariaux (1999).

by the bad meteorological conditions, besides a deterioration of the cell that prevented any calibration back home.

Such an expedition represented an enormous amount of work, was a superb demonstration of ingenuity and of creativity, both instrumental and methodological, but has been frustrated by the too frequent vexations met by time-constrained observations.

4.4. THE CDS

The Centre de Données Stellaires (CDS[29]) has already been described in many places and its two first directors have been presented in Sect. 3. Thus we shall only mention here a few historical facts.

CDS was created in 1972[30], by the French Institut National d'Astronomie et de Géophysique (INAG – see Foonot 19) with the following objectives:
– to compile the most important stellar data available in computer-readable form;
– to improve such data by critical evaluations and comparisons;
– to distribute the results to the astronomical community;
– to carry out its own scientific investigations.
CDS was thus installed at Strasbourg Observatory and managed by a Director answering to a Council of twelve astronomers (including six foreigners).

One of the first tasks of the Center was to establish a huge table of correspondence between the various astronomical catalogs which were then machine-readable. Thus one identification of an object allowed to access all of them as well as all data available in all integrated catalogs. The bibliographical references of the papers where the objects were quoted were made accessible too.

Such a work of modern Benedictines has been the basis of the database SIMBAD that made of CDS a world-wide reference allowing it to be in a strong position for *virtual observatories* projects (see *e.g.* Quinn & Górski 2004, but also Heck 2001). Incidently, the availability of such a tool avoids the recurrence of ludicrous situations of the past where two researchers studied the same star under different names without ever realizing it was the same object.

[29]Renamed in the mid-1980s as Centre de Données astronomiques de Strasbourg to take into account it was also dealing with non-stellar data. For details on current CDS resources, on its services, projects and collaborations, see its web site at `http://cdsweb.u-strasbg.fr/CDS.html` .

[30]See Heck (2005) for the pre-CDS years as well as Jung (2005) on the Origins and early beginnings.

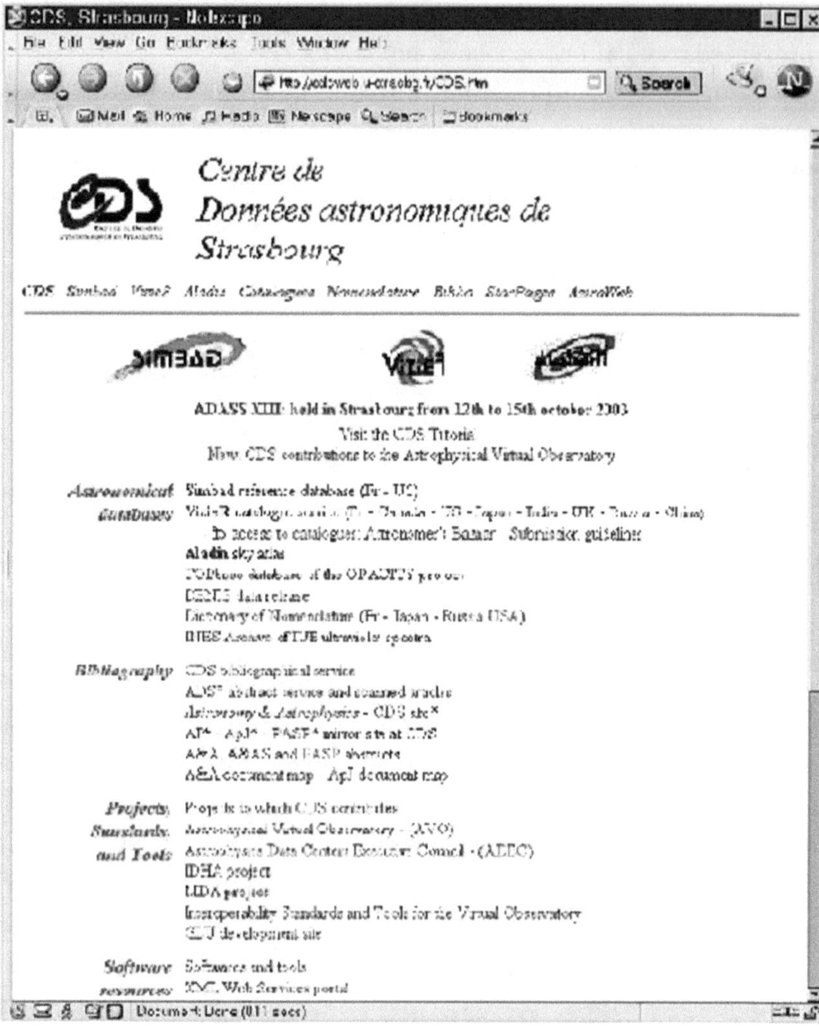

Figure 41. The CDS web page.

4.5. THE HIPPARCOS SATELLITE

As pointed out by Kovalevsky (2005), the genesis of the space astrometric mission Hipparcos took place essentially at Strasbourg Observatory. Around 1965, Lacroute envisaged pointing stars from spacecraft – a daring approach at a time where space astronomy was only in its infancy.

Lacroute's proposal – an idea of genius since it is applied now for all new projects of astrometric satellites – was to use a complex mirror pointing stellar fields with a constant reference angle. A grid positioned in the tele-

Figure 42. Pierre Bacchus and Pierre Lacroute at the Aussois Hipparcos colloquium in June 1985. (© C. Turon)

scope focal plane screened the photon fluxes while the satellite scanned the sky. The analysis of light modulations from two stars enabled determining their angular distance.

A first draft of the project was sent in 1966 to the French space agency, the Centre National d'Études Spatiales (CNES), and then evolved towards an ever more elaborated version. As the French space program had been discontinued in 1971, a new proposal was sent to the European Space Research Organization (ESRO) in 1973. Lacroute, with the collaboration of Pierre Bacchus (in Strasbourg, then in Lille after 1973 – Fig. 42), tackled also the challenge of the data reduction.

Accepted as a program by the European Space Agency (ESA, succeeding ESRO) in 1980, Hipparcos was launched on 8 August 1989. The data collected during the operational life of the satellite (ended on 15 August 1993)

Figure 43. The Hipparcos satellite, born in Strasbourg. (© ESA)

led to very numerous investigations as demonstrated by the proceedings[31] of the impressive scientific colloquium organized in Venice by ESA in 1997. Lacroute had passed away in 1993 and could not see the enormous catalog resulting from his brainchild.

5. More Great Men ...

This section will be devoted to a few prominent scientists who have been associated with Strasbourg Observatory. It is of course out of the scope of this chapter to mention here all astronomers who contributed to the notoriety of the institution. A detailed list of personnel compiled from the observatory publications and archives, and as complete as possible, can be found elsewhere in this volume.

[31] Proceedings published under the reference ESA-SP 402.

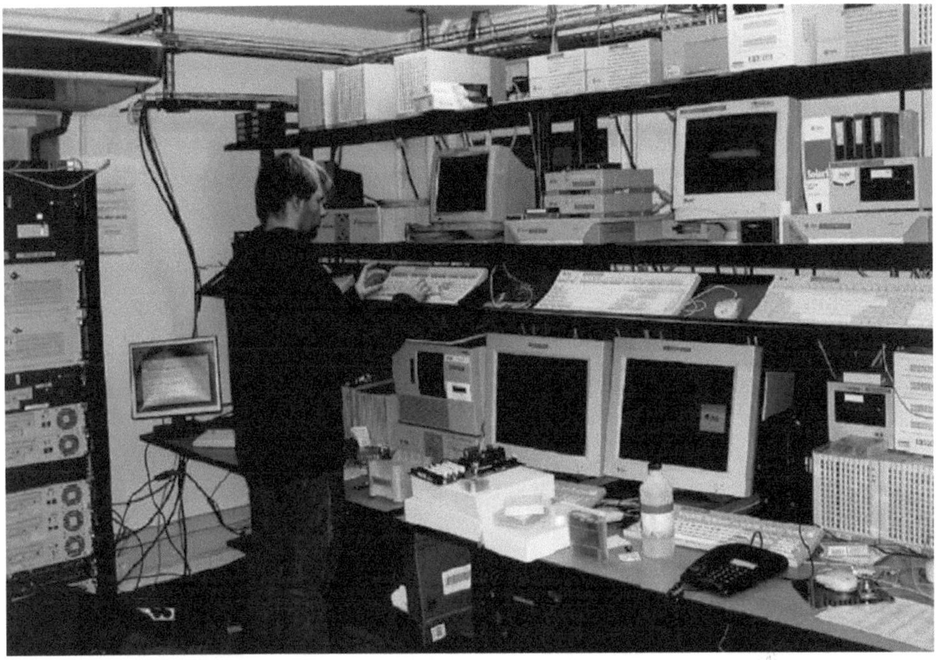

Figure 44. Tools of today: one of the observatory rooms full of computers and telecommunications/networking machines. (© A. Heck)

5.1. THE FIRST GERMAN PERIOD

5.1.1. *W.F. Wislicenus*

Contrary to what many people tend to believe today, extensive bibliographic sources were not born with computers. Beyond a few undertakings dating from the end of the 17th century, Walter (Friedrich) Wislicenus (1859-1905) initiated at Strasbourg an exhaustive compilation that lasted one century.

From the life and career of Wislicenus[32], let us retain here that he was born in Halberstadt (Saxe-Anhalt) and went to schools in Berlin and Dresden. He then tackled studies in mathematics and astronomy at the University of Leipzig. Wislicenus became *Assistent* at Strasbourg Observatory in 1884, then *Privatdozent* [33] at the University in 1889, and finally *Außenordentlicher Professor* (Extraordinary Professor) in 1894.

[32]For more details, see the obituary by Kobold (1906) and the chapter by Duerbeck (2005c) in this volume.

[33]His title today would probably be *Lecturer* or *Reader*, without being formally paid by the University.

Figure 45. Walter Friedrich Wislicenus (1859-1905). (© Astron. Gesellschaft)

Under the patronage of the *Astronomische Gesellschaft*, he published, from 1899 and until his death, the *Astronomischer Jahresbericht* (AJB), an annual bibliographical compendium that survived him. The 68th and last volume under that name was published in 1969[34] (literature of the year 1968) by the Astronomisches Rechen-Institut in Heidelberg. But the series went on under the title *Astronomy and Astrophysics Abstracts* (A&AA) until 2001 (literature of the year 2000) when it was discontinued indefinitely. It was indeed difficult to stand competition with resources such as the *Astrophysics Data System* (ADS) available on the Internet[35].

5.1.2. *C.W. Wirtz (and H.O. Rosenberg)*

Born in Krefeld (Rhineland), Carl (Wilhelm) Wirtz (1876-1939) studied at Bonn Observatory. After staying a while in Hamburg, he joined Strasbourg Observatory in 1902. There he observed essentially nebulae and double stars with the Large Refractor, but his measurement for the diameter of Neptune in 1903 has remained a reference during decades (Seitter & Duerbeck 1999).

Duerbeck & Seitter (2005) detail Wirtz's life and research, while Theis *et al.* (1999) rather analyze the events of the national-socialist period. Wirtz

[34]That year 1969 saw also the merging (with the financial support of ESO) of several national European professional journals into a single one under the title *Astronomy and Astrophysics* (A&A).

[35]`http://adswww.harvard.edu/`

Astronomischer Jahresbericht

Mit Unterstützung der

Astronomischen Gesellschaft

herausgegeben von

Walter F. Wislicenus.

I. Band

enthaltend

die Litteratur des Jahres

1899.

Berlin.

Druck und Verlag von Georg Reimer.

Figure 46. Cover of the first volume (1899) of the *Astronomischer Jahresbericht (AJB)* edited by Wislicenus, with the stamp of Strasbourg Imperial Observatory. (scan by the author)

has been definitely the most active and most productive astronomer of the German times in Strasbourg. His work was critically disturbed by WWI and his personal life was also seriously hampered by the rise of national-socialism. He died just before WWII.

Wirtz had left Strasbourg in September 1916 when he was assigned to the war headquarters in Berlin. He was then busy with geodesy, astronomy, cartography and ballistics, as well as with some teaching. He had been nominated Professor in Strasbourg in 1909, at a time when the directorship was passing from Becker to Bauschinger (see Sect. 3).

Four years earlier, Wirtz had married Helene Borchardt whose sister Verena had just married Hans (Oswald) Rosenberg (1879-1940) who was originally from Berlin and studying then in Strasbourg. Wirtz and Rosenberg were sharing a strong interest in photometry. After some wandering

Figure 47. Carl Wilhelm Wirtz (1876-1939) around 1903 (left) and around 1930 (right). (© Astron. Gesellschaft)

due to the troubled period and a stay at Yerkes Observatory, Rosenberg became Professor and Director of Istanbul Observatory where he died of a heatstroke (Theis *et al.* 1999).

After WWI, Wirtz went to Kiel Observatory directed then by Harzer. There were also, not only Kobold (earlier in Strasbourg too – see Sect. 3) who was editing the *Astronomische Nachrichten*, but also Hellerich, then assistant, who was going to be put in charge of the observatory during WWII.

Wirtz has been occasionally called the *Hubble without telescope* with reference to the lack of support and means he experienced in Kiel. He studied various relationships between the physical parameters of "nebulae" (Duerbeck & Seitter 2005), but, as many precursors in cosmology of the time, he remained largely ignored.

Had he the opportunity to read before his death *The Realm of Nebulae* published by Hubble in 1936? That book, translated into German (Hubble 1938) by Karl-Otto Kiepenheuer from the University of Göttingen, is making reference to Wirtz's investigations.

It should probably be mentioned too that in 1912 (and jointly with Kobold) Wirtz received the Lalande Prize from the French Academy of

Figure 48. André Couder (1897-1979). (© Acad. Sciences Inst. France)

Sciences[36]. According to Theis *et al.* (1999), Wirtz would have declared that the arrival of French troops in Strasbourg at the end of WWI marked the happiest day of his life. Does this partly explain why Kobold denounced him a few years later during the national-socialist regime?

5.2. THE FRENCH PERIODS

5.2.1. *A. Couder (and G. Rougier)*

In 1925, Strasbourg Observatory has been a launching platform for a young brillant chemist who was going to be associated during half a century to the development of world-class astronomical optics.

Born in Alençon (Orne), engineer graduated in 1919 of the École de Chimie de Paris, André Couder (1897-1979) constructed his first refractor at the age of 14. After two years in the industry, he got a position as auxiliary assistant at the Strasbourg Institute of Chemistry under the direction of Louis Hackspill. But if the days were chemical, the nights were astronomical at the observatory.

With Gilbert Rougier[37], Couder was preparing the first photoelectric

[36]See *C.R. Acad. Sciences Paris* **255** (1912) 1302-1303.

[37]Born in 1886 in La Mulatière (Rhône), Rougier joined Strasbourg Observatory at the end of 1919. He became Director of Bordeaux Observatory in 1937. He had considered

Figure 49. Gilbert Rougier (1886-1947). (© Obs. Astron. Bordeaux)

cells in Hackspill's laboratory. At the observatory, he was involved with meridian observations. He was also studying elastical and thermical distortions limiting the precision of measurements by modifying instrumental characteristics (Fehrenbach 1979).

Couder was nominated *Assistant Stagiaire* at Strasbourg Observatory in 1925. His CV (Couder 1954) mentions also observing campaigns in Southern France during the Summers of 1924 and 1925 organized by Danjon at the request of Esclangon. The site for what would become Haute Provence Observatory was already surveyed. In September 1925, Couder was transferred, on the proposal of Danjon and General Ferrié, to Paris Observatory where he joined the Optical Laboratory where was then working the renowned American optician G.W. Ritchey. In July 1926, Couder was put in charge of the laboratory and, from then on, optics in all its acceptions monopolized his attention and brought him a world-wide reputation together with attractive foreign offers[38] that he declined.

applying for the directorship of Quito Observatory in Ecuador. For more details, see Danjon (1947) and Véron (2004). See also Heck (2005) on Danjon & Rougier's observations of the green flash.

[38]Such as, in 1924, an offer from Otto Struve to polish mirrors for large American telescopes. See Fehrenbach (1990) for more details on this as well as on the genesis of the Parisian optical laboratory. Otto Struve belongs to the fourth generation of the Struve dynasty (see Footnote 12).

As mentioned earlier (see Sect. 4), he had modified the objective lenses of the double equatorial used by the Strasbourg group for the eclipse expedition in Poulo Condore in 1929. In 1933, he tackled the objective lenses of the Strasbourg Large Refractor, significantly improving the instrument performance (Couder 1936).

The volume *Lunettes et Télescopes* (Danjon & Couder 1935) has been mentioned earlier (Sect. 3). Levy (2003) was commenting the occasional irritation of Danjon calling Couder a *cossard* (lazy fellow) as he did not write down his part fast enough!

5.2.2. *A. Lallemand*

Born in Cirey (Côte d'Or), André Lallemand (1904-1978) studied at the University of Strasbourg. After a year spent in high-school teaching, and in spite of other offers, Lallemand accepted in 1928 a position as *Aide-Astronome* at Strasbourg Observatory. He knew the place as he had been helping out there as an assistant from 1925 to 1927. He was promoted as *Astronome-Adjoint* in 1938, still in Strasbourg before being officially assigned to Paris in 1943.

Barely associated with Strasbourg Observatory, Lallemand actively took his share in preparing the total solar eclipse expedition to Poulo Condore (Sect. 4). There he obtained the first infrared photographs of the solar corona. His microphotometric measurements confirmed the existence of the white corona that he then correctly described as a plasma.

Lallemand's instrumental *virtuosity* (Danjon 1960b) led him to propose as early as 1933 an instrumentation based on the photoelectric effect and aiming at reducing the long exposure times required by faint astronomical objects. The *electronic camera* [39] was thus taking shape and it would soon be called the *Lallemand camera.*

Interrupted by WWII, the trials resumed in 1949 and satisfactory electronic photographs were obtained during the 1950s – results then recognized and appreciated abroad as testified by the following flattering citation by Gerald E. Kron (1959), himself an inventor of largely used amplyfing tubes:

> *"Use of an image tube can offer three advantages over direct recording with a photographic plate:*
> *(1) greater speed, by a factor of 50 to 100;*
> *(2) relative freedom from grain;*
> *(3) linearity between blackening and light intensity.*
> *"All three of these advantages have been demonstrated by A. Lallemand and his co-workers, working with the Lallemand image tube.*

[39] The term used then was *electronic telescope* in a parallel with the electronic microscope that had just proved its capabilities.

Figure 50. André Lallemand (1904-1978) in a picture taken at Paris Observatory around 1950. (© Acad. Sciences Inst. France)

> *"At the moment the Lallemand type of image tube seems to be the only one to offer to their fullest extent all the advantages just listed."*

5.2.3. *P. Muller*

Paul Muller (1910-2000) is the only person mentioned here who has been durably at Strasbourg Observatory before and after WWII. Born in Lorquin (Moselle), Muller joined the observatory in 1931 where Danjon asked him to study the astronomical applications of birefringent quartz prisms. Prisoner during all WWII, Muller had to wait 1948 to defend his thesis on a novel double-image micrometer[40].

Muller then devoted his studies essentially to double stars where *"his micrometer demonstrated a precision and a convenience of usage much superior to those of wire micrometers"* (Bacchus 2005). Muller left Strasbourg in 1956 for Paris Observatory. For his observations, he was a regular user of the two largest French refractors (Paris and Nice). He also monitored artificial satellites from the first launches, in particular with cameras installed for this purpose in Strasbourg (see Sect. 4).

See also Débarbat (2005) on Lallemand and Muller.

[40]Such a micrometer can thus be considered as resulting from Danjon's ideas on the double-image principle, an ancient principle (heliometer) brought to Danjon's attention by Couder (Levy 2003). Muller's thesis was presented in Paris.

Figure 51. Paul Muller (1910-2000) with his first refractor at the age of 17. (© Ch. Muller)

5.2.4. *J. Delhaye (and others)*

Though he has never been a member of Strasbourg personnel, Jean Delhaye (1921-2001) deserves however a few words here, so big was his influence on the recent life of the institution[41].

Born in Lourches (Nord), Director of Besançon Observatory from 1957 to 1964, Deputy Director of Paris Observatory from 1964 to 1968 (Jean-François Denisse having succeeded Danjon as Director), then Director from 1968 to 1971, later Director of INAG from 1971 to 1979, Delhaye is CDS' spiritual father. He indeed set it up and installed it in Strasbourg with the support of French and foreign colleagues who understood the appropriateness of the project.

A sample of these can be found as members of the first CDS Council: A. Bijaoui (Nice), A. Blaauw (ESO), J. Boulon (Paris), G. Cayrel de Strobel (Meudon), Ch. Fehrenbach (Haute Provence), W. Fricke (Heidelberg), B. Hauck (Lausanne), C. Jaschek (La Plata), G. Larsson-Leander (Lund) & C.A. Murray (Greenwich), besides Delhaye himself and, *ex officio*, J. Jung as first CDS Director.

[41] In his CV (1979), Delhaye is in fact giving very little space to CDS, merely listed at the same level as other *INAG services of general interest* such as the computer center at Meudon Observatory, the Centre de Dépouillement des Clichés Astronomiques (CDCA) at Nice Observatory or a small plane for atmospheric research.

Figure 52. Jean Delhaye (1921-2001) together with Catherine Turon at a 1995 meeting in the observatory. (Phot. J. Marcout, © Obs. Astron. Strasbourg)

Fig. 52 shows Delhaye together with Catherine Turon (Paris) who played a critical rôle in the success of the Hipparcos project (see Sect. 4), as well as in the exploitation of the corresponding data since she headed the INCA consortium, one of the two structures in charge of reducing the data collected by the spacecraft.

6. Epilogue

A French national champion of field athletics from just before WWII was recently telling his story. During that conflict, he wore first on the Eastern front the German uniform of the some 130,000 Alsatians and 30,000 Mosellans *"malgrés nous"* (enlisted in spite of themselves) and under which they were considered as little reliable by the nazis.

Prisoner of the Red Army, he knew the harshness of Tambov concentration camp[42] before being liberated by drawing. He then switched from the Soviet uniform to the British one while transiting in Tehran, then to American clothes when joining in Algiers Général Leclerc's '2e DB' group. He was finally demobilized under the French uniform after reduction of

[42]That camp (n° 188) was located at about 450km Southeast of Moscow. One estimates that some 17,000 Alsatians and Mosellans died in Soviet camps.

Figure 53. The current Big Dome mirrored in the nearby pool of the Botanic Garden (cf. Fig. 15). (Phot. J. Marcout, © Obs. Astron. Strasbourg)

Adolf Hilter's so-called Eagle Nest[43] in the Berchtesgadener Alps. One of his friends went via the Baltic Sea, then down through the Benelux where he was enlisted as an interpreter by the British troops moving towards Berlin. Such itineraries are far from being isolated cases.

This chapter started with a description of the Alsatian context. It is

[43]The original nickname *Nid d'Aigle* was given by French Ambassador André François-Poncet when reporting a meeting with the Reich Chancellor in that place.

ending with this example of the wounds[44] experienced by the region through its recent history, some of which are still very sensitive. At a time when, since Carolus Magnus, Europe is uniting again and regions are recovering a rising identity, sometimes through national borders, Strasbourg (rather than Alsace) is seen today in a European perspective. Quite a number of international organizations are indeed based there now.

For the period covered by this chapter, we saw that Strasbourg Observatory and its personnel were essentially non-Alsatian. Studies have been devoted to the nationalistic rôle of academic institutions in that region contested by two countries, as well as to the position of teachers and students towards the local Alsatian society[45]. For Strasbourg Observatory, such considerations should probably by modulated by the open mind, the tolerance and the pragmatism generally shown by astronomers inspired by the cosmic character of their science and the international, sometimes planetary, collaborations characterizing it.

Since the end of WWII, political conditions in Europe have dramatically evolved: European integration process initiated by the 1957 Treaty of Rome[46], Franco-German reconciliation ratified by the 1963 Élysée Treaty, and so on. Mentalities spectacularly changed too, influenced by the progress in education, in telecommunications, in transportation means, etc., as well as by the omnipresence of medias and by the mass travel phenomena, not to forget the now common view from space of our planet, mankind's ship, inducing the notion of "small world" with all possible connotations in peoples' subconscient.

The statutes regulating research institutions have also significantly evolved over the last decades. The universities have seen their student population exploding, generating enormous teaching requirements tapping human resources often to the prejudice of other initial fundamental missions such as service, research and knowledge progress. Entities created with a specific identity, the French observatories have become research units with a derogatory status (*i.e.* hybrid with a direct link towards the ministry and the other one towards the local university). They are occasionally under pressure either to become laboratories purely associated to the French Centre National de la Recherche Scientifique (CNRS – National Center for Scientific Research) or to be integrated within more important university centers meaning, in either case, the end of their own existence.

[44] A synthesis of these wounds, certainly romanced but well targeted, can be found in the TV series entitled *Les Alsaciens ou les Deux Mathilde* (Arte/FR3 Video 1996).

[45] See for instance the essay by Craig (1984) for the period 1870-1939.

[46] Some historians go back to the 1944 Benelux Treaty (signed in London by the exiled government of the three countries involved), sometimes even to the 1924 Monetary Agreement between Belgium and Luxembourg ...

Nowadays the organigram of Strasbourg Observatory is rather complex with people depending from various administrations: Comité National des Astronomes et Physiciens (CNAP – National Committee of Astronomers and Physicists), CNRS, universities, not to forget specific contracts (space agencies, European projects, international collaborations, etc.). The current staff is varied and cosmopolitan, a sign of the times certainly, but also an indication that the institution succeeded in making itself attractive while reaching a world-class excellence in the course of its short history.

Acknowledgments

It is a real pleasure to acknowledge here the help of the many persons who assisted us in gathering together pieces of information and illustrations presented in this chapter. Our warmest thanks go more particularly to Mmes & Messrs P. Abraham, Ch. Bruneau, N. Cramer, P. Dubois, H.W. Duerbeck, A. Fresneau, J. Garnier Fl. Greffe, M. Haller, M. Hamm, J. Jung, Th. Keller, J. Kovalevsky, J. Krautter, Fr. & J.M. Lacroute, S. Langenbacher, J. Levy, L. Ludes, L. Maison, J. Marcout, B. Moya, Ch. Muller, C. Schohn, E. Schweitzer, W. Seggewiss, W.C. Seitter, M. Sessa, B. Traut, C. Turon, M. & Ph. Véron, Ph. Vonflie, F. Woelfel, G. Wolfschmidt, as well as the archivists of the Académie des Sciences de l'Institut de France.

References

1. Bacchus, P. 2005, Paul Muller (1910-2000), this volume.
2. Becker, E. 1896a, *Ann. Kaiserl. Univ.-Sternw. Strassburg*, 1. Serie, **1**, I-XVII.
3. Becker, E. (Ed.) 1896b, *Ann. Kaiserl. Univ.-Sternw. Strassburg*, 1. Serie, **1**, XCVIII + 350 pp.
4. Comerón, F. 2004, Observing in Service Mode: The Experience at the European Southern Observatory, in *Organizations and Strategies in Astronomy 5*, Ed. A. Heck, Kluwer Acad. Publ., Dordrecht, 141-158.
5. Couder, A. 1936, L'Objectif du Grand Réfracteur de Strasbourg – Retouche et Nouvelle Étude – Compensation de la Biréfringence, *Ann. Obs. Strasbourg* **III**, 113-134.
6. Couder, A. 1954, Notice sur les Titres et Travaux Scientifiques, Archives Acad. Sciences, Paris.
7. Craig, J.E. 1984, Scholarship and Nation Building – The Universities of Strasbourg and the Alsatian Society 1870-1939, Univ. Chicago Press, xii + 516 pp. (ISBN 0-226-11670-0)
8. Danjon, A. 1947, Gilbert Rougier, *Bull. Soc. Astron. France* **61**, 143-145.
9. Danjon, A. 1958, The Contribution of the Impersonal Astrolabe to Fundamental Astronomy, *Monthly Not. Roy. Astron. Soc.* **118**, 411-431.
10. Danjon, A. 1960a, The Impersonal Astrolabe, in *Telescopes*, Eds. G.P. Kuiper & B.M. Middlehurst, Univ. Chicago Press, 115-137.
11. Danjon, A. 1960b, Rapport sur les Titres de M. André Lallemand, Archives Acad. Sciences, Paris.
12. Danjon, A. & Couder, A. 1935, Lunettes et Télescopes, Blanchard, Paris, 712 pp.
13. Danjon, A., Lallemand, A. & Rougier, G. 1938, Éclipse Totale de Soleil du 9 Mai

1929 – Mission de l'Observatoire de Strasbourg à Poulo Condore, *Ann. Bureau Long.* **11**, B1-B36.
14. Débarbat, S. 2005, Strasbourg Observatory: A Breeding Place for French Astronomical Instrumentation in the 20th Century, this volume.
15. Delhaye, J. 1979, Titres et Travaux, Archives Acad. Sciences, Paris.
16. Demariaux, M. 1997, Poulo-Condore, Archipel du Viêtnam, L'Harmattan, Paris, 265 pp. (ISBN 2738482546)
17. Duerbeck, H. 2004, Venusdurchgänge zu Kaisers Zeiten, *Sterne u. Weltraum* **43**, 34-42.
18. Duerbeck, H. 2005a, The Observatory of the Kaiser-Wilhelm University: The People Behind the Documents, this volume.
19. Duerbeck, H. 2005b, Strasbourg Observatory in World War II, this volume.
20. Duerbeck, H.W. 2005c, Walter F. Wislicenus and Modern Astronomical Bibliography, this volume.
21. Duerbeck, H.W. & Seitter, W.C. 2005, The Nebular Research of Carl Wirtz, this volume.
22. Esclangon, E. 1926, La Nouvelle Organisation de l'Observatoire, *Ann. Obs. Strasbourg* **I**, 1-44.
23. Esclangon, E. 1929, Rapport de M. Esclangon sur M. Danjon, Archives Acad. Sciences, Paris.
24. Esclangon, E. 1930, Titres et Travaux Scientifiques, Presses Univ. France, Paris, 42 pp.
25. Esclangon, E. 1945, Rapport sur les Titres et Travaux Scientifiques de M. Danjon, Archives Acad. Sciences, Paris.
26. Fehrenbach, Ch. 1979, Notice Nécrologique sur André Couder, *C.R. Acad. Sciences Paris* **289**, 62-67.
27. Fehrenbach, Ch. 1990, Des Hommes, des Télescopes, des Étoiles, Éditions du CNRS, 528 pp. (ISBN 2-222-04459-6)
28. Hartwig, E. 1898, Friedrich August Theodor Winnecke, *Vierteljahrsschr. Astron. Ges.* **33**, 5-13. – An English translation is available in *Monthly Not. Roy. Astron. Soc.* **58** (1898) 155-159.
29. Heck, A. 2000, Characteristics of Astronomy-Related Organizations, *Astrophys. Sp. Sc.* **274**, 733-783.
30. Heck, A. 2001, Virtual Observatories or Rather Digital Research Facilities?, *Amer. Astron. Soc. Newsl.* **104**, 2.
31. Heck, A. 2002, The Impact of New Media on 20th-Century Astronomy, *Astron. Nahr.* **323**, 542-547.
32. Heck, A. 2005a, Vistas into the CDS Genesis, this volume.
33. Heck, A. 2005b, Strasbourg Green Rays, this volume.
34. Hubble, E. 1936, The Realm of Nebulae, Yale Univ. Press, New Haven, 207 pp.
35. Hubble, E. 1938, Das Reich der Nebel, Fr. Vieweg & Sohn, Braunschweig, xii + 142 pp.
36. Jost, E. 1913, Ernst Becker, *Vierteljahrsschr. Astron. Ges.* **48**, 2-12.
37. Jung, J. 2005, The Stellar Data Center: Origins and Early Beginnings (1972-1974), this volume.
38. Kobold, H. 1906, Walter Wislicenus, *Vierteljahrsschr. Astron. Ges.* **41**, 13-21.
39. Kovalevsky, J. 1967, A Great French Astronomer, *Sky & Tel.* **33**, 347-349.
40. Kovalevsky, J. 2005, The Hipparcos Project at Strasbourg Observatory, this volume.
41. Kron, G.E. 1959, An Image-Tube Experiment at the Lick Observatory, *Publ. Astron. Soc. Pacific* **71**, 386-387.
42. Küstner, F. 1888, Die Sternwarte der Kaiser Wilhelms-Universität zu Straßburg, in *Meyers Konversations-Lexikon* **XV**, Bibl. Inst., Leipzig, pp. 891-894.
43. Lacroute, P. 1956, Le Nouvel Equatorial de 210mm, *Ann. Obs. Strasbourg* **V**, 44-45.
44. Levy, J. 2003, private communication.

45. Mayer, H. 1894, Straßburg und seine Bauten, Herausgeg. Architekten- u. Ingenieurverein f. Elsaß-Lothringen, Straßburg.
46. Moszberger, M., Rieger, Th. & Daul, L. 2002, Dictionnaire historique des rues de Strasbourg, Ed. Le Verger, Illkirch, 430 pp. (ISBN 2-84574-0239).
47. Müller, P. 1975, Sternwarten, Verlag Peter Lang, Frankfurt/Main, 300 pp. (ISBN 3-261-02581-6).
48. Pfau, W. 2000, The Astronomische Gesellschaft: Pieces from its History, in *Organizations and Strategies in Astronomy*, Ed. A. Heck, Kluwer Acad. Publ., Dordrecht, 67-75.
49. Quinn, P.J. & Górski, K.M. (Eds.) 2004, Toward an International Virtual Observatory, Springer, Heidelberg, xxii + 342 pp. (ISBN 3-540-21001-6)
50. Robson, I. 2001, New Strategies for Ground-Based Observing, in *Organizations and Strategies in Astronomy II*, Ed. A. Heck, Kluwer Acad. Publ., Dordrecht, 121-137.
51. Schweitzer, E. 2004, private communication.
52. Seggewiss, W. 2005, Strasbourg Observatory and the Astronomische Gesellschaft, this volume.
53. Seitter, W.C. & Duerbeck, H.W. 1999, Carl Wilhelm Wirtz – Pioneer in Cosmic Dimensions, in *Harmonizing Cosmic Distance Scales in Post-Hipparcos Era*, Eds. D. Egret & A. Heck, *Astron. Soc. Pacific Conf. Series* **167**, 237-242.
54. Stracke, G. 1934, Julius Bauschinger, *Vierteljahrsschr. Astron. Ges.* **69**, 146-163.
55. Strassl, H. 1963, Johannes Hellerich, *Mittl. Astron. Ges.* **17**, 27-30.
56. Theis, Chr., Deiters, St., Einsel, Chr. & Hohmann, F. 1999, Hans Rosenberg und Carl Wirtz – Zwei Kieler Astronomen in der NS-Zeit, *Sterne u. Weltraum* **38**, 127-129.
57. Véron, Ph. 2001, Préhistoire de l'Observatoire de Haute Provence, in *Observatoires et Patrimoine Astronomique Français*, Nantes, 8-9 juin 2001.
58. Véron, Ph. 2004, Rougier, Gilbert (1886-1947), in *Dictionnaire des Astronomes Français*, in preparation.
59. Vonflie, Ph. & Heck, A. 2005, Strasbourg Observatory Institutional Publications, this volume.
60. Winter, L. 2000, Ciel! Mon Mari est muté en Alsace ..., Éd. La Nuée Bleue, Strasbourg, 190 pp. (ISBN 2-7165-0512-8).
61. Wolfschmidt, G. 2005, Strasbourg Observatory in German Times, this volume.

STRASBOURG OBSERVATORY IN GERMAN TIMES

G. WOLFSCHMIDT
Institut für Geschichte der Naturwissenschaften
Fachbereich Mathematik
Universität Hamburg
Bundesstraße 55
D-20146 Hamburg, Germany
wolfschmidt@math.uni-hamburg.de

Abstract. In 1872, Strasbourg University Observatory was founded by the German government as a model observatory – for its construction, its instrumentation as well as its the staff. Unique working conditions were offered. The 49cm refractor was at that time the largest in the German empire and showed the importance given to Strasbourg Observatory. The director and the staff in Strasbourg in the German time from 1872 to 1918 were first-class astronomers. But already in 1872 and again in 1895, one did not succeed to introduce the new topic of astrophysics. Thus Strasbourg became famous for the field of classical astronomy, *i.e.* compiling catalogues of stars with data like positions, magnitudes, proper motions and so on, as well as calculating orbits of planets and comets according to the formulae of celestial mechanics.

1. Introduction

In 1872, after Strasbourg had just become German, the German government decided to found a university which should make visible German culture, education and research, a model university. This *Reichsuniversität*, financed by the Reich (unlike other universities subventioned by the States) and named Kaiser-Wilhelms-Universität, had about 100 professors and was rather big even in comparison to Berlin or Leipzig. The average age of the professors was extremely low, about 38 at the beginning, in comparison to 53 in German universities in general, and 62 in Berlin which was the favourite final goal for many professors.

A. Heck (ed.), The Multinational History of Strasbourg Astronomical Observatory, 63–87.

The university quarter (Mayer 1922) was put up East of the historical center of Strasbourg. This magnificent zone, famous for its town planning, is dominated by a main axis which is in stark contrast to the curved alleys of the old city. This axis, the Kaiser-Wilhelm-Straße[1] leads from the Kaiserplatz[2], in the Northwest to the university in the Southeast. The architectural style of the new university in the reign of Emperor Wilhelm I was considered 'gigantic' by the French (AIV 1894).

The layout of the campus was convenient, with the scientific institutes close to each other. The buildings were designed in collaboration with the directors of the institutes. In the university quarter, the university observatory was visible as a prominent dome in the botanical garden at the end of the axis. Architect Hermann Eggert (1844-1920), specialist in monumental state buildings, had chosen a neo-renaissance style for the building, built from 1877 to 1880[3].

2. Strasbourg as a Model Observatory

Some remarks concerning the tradition of observatory construction are in order. In the 19th century, observatories not only housed the observing instruments and work areas but also contained lodging for the astronomers. A cross was often chosen as a ground plan, *e.g.* for the Athens (1846) and Vienna (1874) observatories: the middle dome housed the main instrument, a refractor. The meridian circle and the transit instrument were in the Eastern and Western observing rooms. Pulkovo Observatory near St. Petersburg, built in 1839, is a fine example of such 19th-century observatory architecture (Struve 1845). Smaller domes for another refractor and a heliometer existed besides the main dome; in between was the meridian circle. This arrangement was also used for Potsdam Astrophysical Observatory, built between 1876 to 1879 (Spieker 1879, p. 33).

In contrast to the above-mentioned observatories, that of Strasbourg consists of three main parts, connected by covered footpaths:
– the refractor building with a prominent dome (Northwest)
– the meridian building between two domes (Northeast) and
– the living quarters and administration building (South).
Architect Eggert was famous for his clearly-arranged and convenient ground plans. The separation of the observing buildings from the study area in the main building and from the apartment house was a completely new idea –

[1]Today: Avenue de la Liberté.

[2]Today: Place de la République

[3]Eggert (1884), p. 79; cf. *Vierteljahrsschrift Astron. Ges.* **14** (1879). In 1889, Eggert was called to head the Ministry for Public Works [Ministerium für Öffentliche Arbeiten] in Berlin; later, the 'model observatory' presented in the Deutsches Museum was that of Strasbourg (Inv.-Nr. 4516, scale 1:50).

Figure 1. Strasbourg Observatory in a 19th-century sketch. (O. Ule, *Die Wunder der Sternenwelt*, Leipzig and Berlin: Otto Spamer (1884), p. 411)

a revolution in observatory construction (Müller 1978, pp. 170-175). This new concept was put in practice for the first time at Strasbourg Observatory. Eggert emphasised that it was the idea of the first director, August Winnecke (according to University Architect Mayer in AIV 1894):

> *"For the arrangement of the total building complex, as well as for the outline of each single building, Professor Winnecke's programme was decisive."*

This construction was the result of Winnecke's experience during his observing activity at Pulkovo observatory. The refractor building is the most important of the observatory and thus has an architecturally dominant dome. It contains the lecture hall, the library and the office of the director. The ground plan of the neo-renaissance building is a Greek cross, and it was built on a single 1.5m-thick concrete slab.

The prominent dome arises above the central octagon. Some empty spaces in the pillars were used for placing the pendulum clocks which thus had a place in the centre of the building and were kept at relatively constant temperature. The dome is 10.5m in diameter, was covered with zinc on wood and was protected against the uneven heating by sunlight by water cooling. Rotation of the dome could be done by hand or electrically. The dome had a elevator floor and a moveable observing chair.

Another observatory building in the East has the L-form typical of small observatories. It has two smaller domes, in addition to the meridian building oriented north-south. The two meridian rooms were erected with an iron framework covered with sheet metal and with wooden jalousies outside which were useful for quick temperature compensation due to their high conductivity. The pillars on which the instruments were mounted were independent of the domes, to avoid vibrations – a technique which was not yet a standard one at that time.

The apartment and administration building had offices and living quarters for the assistants and scientific observers ("Observatoren"); in the upper floor was the flat of the director, in the beginning with a servant's room. In addition, the ground floor served as the porter's lodge. In September 1881, Strasbourg Observatory was inaugurated during the annual meeting of the *Astronomische Gesellschaft*[4] in order to show the building and the instruments to an international public. During this event the significance of the Strasbourg arrangement was emphasized. The president stated (Auwers 1881, p. 262.)

> *"the generally expressed admiration of the astronomers who had the opportunity to see the arrangement and the equipment of the new Strasbourg Observatory."*

The influence of Strasbourg observatory was considerable. Strasbourg became an architectonic model for the Remeis Observatory in Bamberg, built in 1889[5]. Even today, observatories are constructed according to the modern rules invented in Strasbourg: separate foundations for the instruments in the domes, and separation of work, lodging and observing areas. One disadvantage of the observatory was its position in the middle of a city[6]. Already Ernst Becker asked some years after the opening of Lick observatory whether larger observatories should be attached to universities and thus burdened with this local limitations (light pollution, dust, mist from the river), or whether a division should be made into institutions for teaching and observatories devoted to scientific research alone (Becker 1896, p. IX).

3. The most modern instrumentation

The imperial university observatory of Strasbourg was built with a large amount of money. Most of the generous instrumentation was bought from

[4]See the chapters by Bodenmuller & Heck (2005) and by Seggewiss (2005) in this volume (Ed.).

[5]Eggert (1887), pp. 353-354; *Schweizerische Bauzeitung* **10** (1887), p. 80; Hartwig (1887), p. 277 & pp. 329-335.

[6]The next step, building an observatory far outside of a city on the top of a hill or mountain was not accomplished before Nice (1882) or Lick, Mt. Hamilton (1888).

German makers, who were in any case the leading international makers throughout the 19th century[7].

- The main instrument in the dome of the main building is a large 49cm refractor (7m focal length) made by Merz, Munich, 1877, the successor of Fraunhofer. The mounting was made by Oskar Repsold, Hamburg, 1880. This refractor was at that time the largest in the German empire and showed the importance of Strasbourg. Hartwig (1898, p. 12) judged that Strasbourg observatory

  "will always remain a model institution which will keep its efficiency in spite of all improvements which are caused by the rapid progress of technology."

- In the Southern dome was a small 16cm-refractor (6in) made by Reinfelder & Hertel, Munich, 1876, with a micrometer made by Repsold, 1895. The large 16cm comet seeker, Merz, Munich, 1876, mounting made by Repsold, Hamburg, 1880, was also installed here.
- In the Northern dome, a 13.6cm altazimuth made by Merz, Munich, with a mounting made by Repsold, Hamburg, 1879 (improved by Carl Bamberg, Berlin, 1891), was installed.
- A very useful instrument was the heliometer made by Fraunhofer, acquired in 1877 from Gotha observatory. It received a new position circle made by Repsold, Hamburg, 1878. It was put up in a cylindrical building in the garden.
- Small portable refractors and comet seekers:
  - 10.1cm refractor with ring micrometer, Merz, Munich, 1878;
  - 7.6cm refractor, Plössl, Wien, 1878;
  - 4.8cm zenith telescope, M. Hildebrand, Freiberg in Saxony, 1905;
  - 12.2cm comet seeker, Fraunhofer, Munich.
- The 13.2cm transit instrument made by Cauchoix (2m focal length) was taken from the French Academy observatory (1826) in Strasbourg. In 1876 it received a new eyepiece head from Repsold, Hamburg.
- The new meridian circle (16cm diameter) from Oskar Repsold, Hamburg, 1876, was installed in 1880.
- Small portable transit instruments:
  - a new 6.8cm transit instrument, Repsold brothers, 1876;
  - a universal instrument, 1877, for time keeping;
  - a similar transit instrument, Pistor & Martins;
  - a portable 5.4cm transit instrument, Carl Bamberg, Berlin-Friedenau, around 1890;
  - a prism circle, Steinheil, 1878;

[7]Becker (1896); cf. Annual reports of Strasbourg observatory in *Vierteljahrsschrift Astron. Ges.* **12** (1877) to **53** (1918); cf. Schur (1885, 1886).

- a Parisian theodolite, 1878, reconstruction to a spectrometer by the mechanician of the institute;
- a repeating circle, Liebherr, Munich;
- a portable universal instrument, M. Hildebrand, Freiberg in Saxony.

Practically no astrophysical instruments existed – except for a Zöllner photometer made by Ausfeld, Gotha. Only at the time of Julius Bauschinger, who was director from 1909 to 1919, instruments for spectroscopy, photography and photometry were acquired[8].

Finally, the clocks were important for astrometric work. Besides a pendulum clock made by Petit, Paris taken from the Academy observatory, the main clocks in the refractor building were: Hohwü (No. 25) and Knoblich (No. 1963), 1885[9]. In 1907 a Riefler precision pendulum clock (type D) was acquired[10]. Finally a Thomas calculating machine made by A. Burckhardt, Glashütte, 1892, should be mentioned, which was important for carrying out astronomical calculations.

4. Astronomers in Strasbourg from 1872 to 1918

The authorities, wishing to have a first-class staff, tried to hire Karl Friedrich Zöllner (1834-1882), already a famous professor in Leipzig, who had started astrophysics in 1865 and had even introduced the name for this new topic (Zöllner 1865). This would have been a great chance for the development of astrophysics. In 1872, Zöllner got the offer to become full professor in Strasbourg. It was a tempting offer; but after long negotiations with the university of Strasbourg and with the Ministry of Culture of Saxony he declined the offer (Meinel 1991, p. 32):

[8]Later additional equipment for the large 49cm refractor included: a wedge photometer for nebulae and comets, 1908; an actinometer for measuring brightnesses of nebulae made by the mechanic of the observatory, 1910; spectral eyepiece according to Abbe, made by Zeiss of Jena, 1915, for the construction of a spectrophotometer for nebulae. Later additional equipment for the large 16cm comet seeker included: Zöllner photometer made by Ausfeld, Gotha, 1893; 6cm UV triplet for photographic-photometric work and a transmission grating on UV-glass with one groove per mm made by Zeiss of Jena, 1914, for the photographic telescope made in the workshop of the institute.

[9]Further equipment with clocks until 1900 included: regulator with 3/4 second pendulum, contact equipment, Ungerer Frères, 1886, used for the registration with the fourth registration apparatus according to the construction of Fuess, 1889; two new registration clocks, Schweizer, Munich, 1890; anchor escapement precision clock made by Martens, Freiburg im Breisgau; electric dial face (11cm diameter), Hipp, 1894; Hipp chronoscope, 1898; pendulum clock with Riefler pendulum and contact equipment, G. Schmidt-Staub, Karlsruhe, 1898, with registration apparatus made by Fuess, Berlin; precision pendulum clock, S. Riefler, Munich, 1901; chronograph made by Fuess, Berlin, 1905.

[10]Precision pendulum clock (type D) No. 95 with vacuum, free escapement, nickel steel compensation pendulum and electric winder, S. Riefler, Munich (at that time the leading international company for clocks), 1907.

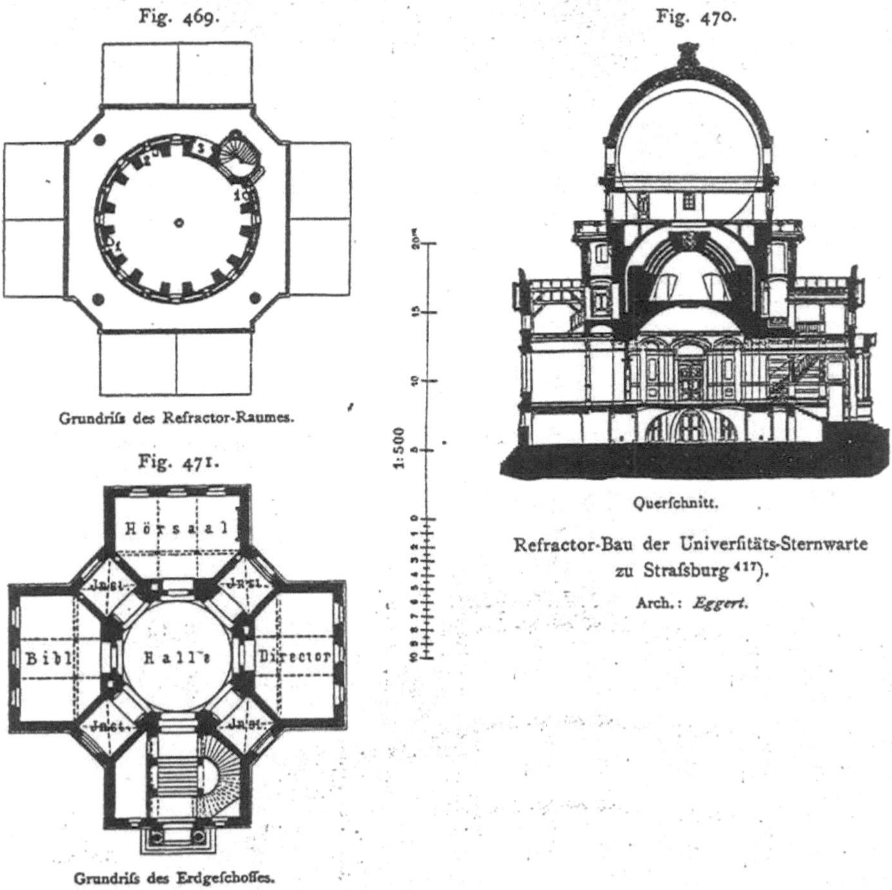

Figure 2. Sketckes of the big dome structure. (Eggert 1884, 82)

> *The sheer thought that someone would call me* Herr Geheimrath *[Mr. privy counciller] instead of professor, or would bestow a decoration as a recognition for my previous achievements aroused, because of the presently low rate of such equivalents, aroused such a painful feeling within myself, that I will hopefully be protected for all times from such decorations, regardless of their provenience, by this public confession.*

Zöllner did not want to have such a socially prestigious position because he was very modest and liked to be in the background. Thus he decided to remain in Leipzig as a full professor in a newly-created position of *physical astronomy*[11].

[11] Zöllner, Karl Friedrich: Personalakte im Universitätsarchiv Leipzig, p. 4. I am grateful to Gisela Münzel, Leipzig, for this information (1995). Koerber (1899), p. 68; Zöllner

TABLE 1. Staff in Strasbourg – Winnecke Era

Position	Year	Name	Notes
Dir	1872-1886	Winnecke, A.	"on leave" 1885-86
(Dir)	1882-1886	Schur, W.	1886 Dir Goettingen
(Dir)	1886-1887	Kobold, H.A.	
Obs	1877-1886	Schur, W.	
Obs	1886-	Kobold, H.A.	
Asst	1873-1877	Schur, W.	
Asst	1877-1884	Hartwig, E.	later Obs Dorpat, Dir Bamberg
Asst	1884-1889	Wislicenus, W.F.	
2-Asst	1882-1884	Wislicenus, W.F.	
2-Asst	1884-1885	Stechert, C.F.	later Prof Seewarte hamburg
2-Asst	1884-1889	Kaufmann, A.	later secondary school teacher
Servant	1873-1881	Kunkelmann	
Servant	1882-1884	Klemann	
Servant	1884-1886	Brauns, L.	
Servant	1886-1888	Oelke, H.	
Porter	1881-1888	Sabel, C.	

Abbreviations: Dir = Director, (Dir) = Provisional Director, Prof = Professor, Obs = Observer, Asst = Assistant, 2-Asst = second Assistant

Due to this refusal, and the appointment as Director of Winnecke, an astronomer with great experience in positional astronomy, Strasbourg became a model observatory of classical astronomy, suitable for compiling large star catalogues of high precision, a field of research with a long tradition in Germany.

August Winnecke[12] (1835-1897), son of a pastor, studied in Göttingen and Berlin universities and then received further training with Friedrich Wilhelm Argelander (1799-1875) in Bonn (1856), and in Pulkovo observatory in St. Petersburg with Friedrich Georg Wilhelm von Struve (1793-1864). In 1864, he married Hedwig Dell, the niece of Otto Wilhelm von Struve (1819-1905), then director in Pulkovo. Winnecke even became Vice-Director in Pulkovo at the age of 30. He was appointed Director in Strasbourg in 1872 and, in 1882, he was even chosen to be Vice-Chancellor of the university. While in Strasbourg, Winnecke received prestigious offers, once in 1875 to succeed Argelander (1799–1875) in Bonn, another time in 1881 to be the successor of Johann von Lamont (1805–1879) in Munich, but

(1881), p. 815.

[12]Hartwig (1898), Foerster (1898).

he declined them. Compared to Strasbourg, neither of these observatories offered better instruments nor a better budget.

The whole staff including the servant, the mechanic and the porter, were not from Alsace, but from Germany. Winnecke's first staff members (cf. Table 1) were the assistant, and later observer, Wilhelm Schur and the assistant Ernst Hartwig.

Wilhelm Schur[13] (1846-1901), born in Altona, was related to August Christian Petersen (1806-1880), the director of Altona Observatory, and hence had an early interest for astronomy. He completed his degree in Kiel and Göttingen in 1867 with a dissertation about orbits of double stars. In 1868 he became assistant at the Geodetic Institute in Berlin, the *Centralbureau der Europäischen Gradmessung*. He began working in Berlin in 1872, and in 1873 in Strasbourg, preparing the German Venus transit expedition; a highlight was his participation in the expedition to the Auckland Islands in 1874. In 1881 he wrote his habilitation thesis in Strasbourg with the topic *Determination of the mass of Jupiter using accurate heliometer measurements of the moons of Jupiter*. From 1882 to 1886 he was acting director in Strasbourg. In 1886, he became professor and director of Göttingen Observatory.

Ernst Hartwig[14] (1851-1923), born in Frankfurt, passed the *Abitur*[15] at the famous Melanchthon-Gymnasium in Nuremberg. He studied in Erlangen, Leipzig, Göttingen and Munich. He gained his astronomical experience, especially the observation of variable stars, with Winnecke in Strasbourg, starting in 1874. In 1884, he became observer in Dorpat (today Tartu, Estonia). In addition to several study trips, he led the German Venus transit expedition to Bahia Blanca, Argentina, in 1882. In 1886 he was invited to act as the first director of the Remeis Observatory in Bamberg. In 1916 he became honorary professor at Erlangen University.

After the scientific observers Schur and Kobold had acted as substitutes for Winnecke, who had been ill since 1882, the Faculty for Mathematics and Science requested that the ministry should find a new director (Jost 1913, p. 7):

> *When Strasbourg University was newly installed, the construction an observatory was funded by large means, and Winnecke was given the task to create an institute whose perfection of equipment was unique within Germany... But a harsh fate denied Winnecke the harvest of his seed, when he succumbed to the illness that soon forced him into retirement, so that the observatory was deprived of its director and creator less than*

[13]Brendel (1902), Meyermann (1902).

[14]Hoffmeister (1923), Heise (1923).

[15]Diploma from a German secondary school or 'Gymnasium' qualifying for university admission.

half a year after its completion. It was not easy to find a replacement, and with the hope for his recovery, the director's post was not filled definitively.

The first choice of the faculty was Hugo von Seeliger (1849-1924), who had the chair in Munich as full professor and director since 1882:

Herr Seeliger steht etwa in der Mitte der dreißger Jahre. Derselbe war früher Director der Sternwarte Gotha und wurde von dort an die Universität München [...] berufen. Seine Arbeiten erstrecken sich über die verschiedenen Gebiete der Astronomie; er ist gleichzeitig hervorragender Theoretiker wie praktischer Astronom und in der Leitung von Sternwarten erfahren.[16]

[Mr. Seeliger is in his mid-thirties. He was previously director of Gotha Observatory, and was then called to Munich University[...] His studies comprise various parts of astronomy; he is at the same time an excellent theoretician and a practical astronomer and is experienced in directing an observatory.]

After hesitating for a long time, Seeliger finally declined the tempting offer:

Die Straßburger Sternwarte ist so überaus schön, ja großartig, eingerichtet, daß ich sie für das beste und zweckmäßigste Institut in ganz Deutschland und Österreich halte und es mußte mir außerordentlich schwer fallen, darauf Verzicht zu leisten, einen solchen Schatz für meine wissenschaftlichen Arbeiten nutzbar zu machen.[17]

[Strasbourg Observatory is most beautifully, extraordinarily equipped, that I consider it to be the best and most useful institute in all of Germany and Austria. I found it most difficult to decline the invitation to make use of such a treasure for my scientific work.]

Preferred candidates were the most prominent astronomers of the day who already had a position as a director in addition to a professorship. But this caused also a number of rejections in 1887[18]:

– Friedrich Tietjen (1832-1895), since 1874 director of the *Astronomisches Recheninstitut* in Berlin;

– Eduard Schönfeld (1828-1891), since 1859 director in Mannheim, since 1875 director in Bonn (he declined due to health problems); he wanted to study the blue prints of the observatory and he suggested improvements in organization; *e.g.* he asked for a better salary for the assistants and observers and requested more money for publications;

– Heinrich Ernst Bruns (1848-1919), since 1882 director in Leipzig (he had already declined the Göttingen directorship in 1885).

[16]Eingabe der mathematisch-naturwissenschaftlichen Fakultät ans Ministerium vom 8. März 1886. Arch. Dép. Bas-Rhin, AL 103, pag. 71, No. 245, Bl. 34. cf. Berufungsvorgang Bl. 34-41.

[17]Seeliger, Hugo von: Schreiben an die Universität Straßburg, Unterstaatssekretär Ledderhose, 13.6.1886. Arch. Dép. Bas-Rhin, AL 103, pag. 71, No. 245, Bl. 40.

[18]Arch. Dép. Bas-Rhin, AL 103, pag. 71, No. 245, Bl. 42-71.

Now, Schönfeld suggested especially Krüger, Peters and Ernst Becker as director for Strasbourg; as well as and Hermann Struve:
– Adalbert Krüger (1832-1896), from 1862 director in Helsingfors, from 1876 director in Gotha, from 1880 director in Kiel;
– Carl Friedrich Wilhelm Peters (1844-1894), in 1873 appointed director in Kiel (in 1888 in Königsberg);
– Hermann Struve (1854-1920), astronomer in St. Petersburg.

Wilhelm Foerster suggested Rudolf Lehmann-Filhés (1854-1914), who was since 1881 private lecturer ('Privatdozent') at Berlin University (since 1891, professor in Berlin).

Arthur von Auwers (1838-1915), astronomer at the Prussian Academy of Sciences, suggested Ernst Becker; finally this proposal was accepted and Becker was offered the directorship[19].

During his study of mathematics, Ernst Becker (1843-1912) already had contacts with Wilhelm Foerster (1832-1921), director of the Berlin observatory, and with Friedrich Tietjen (1832-1895), who recognized his great mathematical ability relatively early. After a dissertation dealing with the calculation of perturbations, Becker became observer in Leiden in 1870, in Neuchâtel, Switzerland in 1871, and in Berlin in 1874. In 1883, he became director of Gotha Observatory.

Among Becker's staff (cf. Table 2) in the 1880s was Walter Friedrich Wislicenus and Hermann Albert Kobold and later, Carl Wilhelm Wirtz.

- Hermann Albert Kobold[20] (1858-1942) finished his studies in Göttingen in 1880 with a dissertation on the calculation of perturbations. During his position as an assistant in O'Gyalla, Hungary, he became acquainted with astrophysics, especially with cometary spectra. In 1882 he participated in the German Venus expedition to Aiken/South Carolina, and afterwards he was a member of the team in Berlin that analysed the observations. In 1887 he became observer in Strasbourg. After his habilitation in 1888, he was named in 1900 'außerordentlicher Professor', a title which carried no salary. In 1902 he got the paid position of an observer and assistant professor in Kiel. He became the editor of *Astronomische Nachrichten*, a leading international astronomical journal, a position he kept until 1938. In 1909 – after declining a call to Berlin – he was appointed honorary professor, and in 1921 full professor in Kiel.
- Walter Friedrich Wislicenus (1859-1905) descended from a family of liberal-minded Protestants[21]. He studied mathematics and astronomy at the university of Leipzig, where Zöllner and Seeliger were among his

[19]Arch. Dép. Bas-Rhin, AL 103, pag. 71, No. 245, Bl. 72-76.
[20]Ebell (1942).
[21]Kobold (1906), p. 13.

TABLE 2. Staff in Strasbourg – Becker Era

Position	Year	Name	Notes
Dir	1887-1909	Becker, E.	
Obs	1887-1902	Kobold, H.A.	later Obs Kiel
Obs.	1902-1918	Wirtz, C.W.	
Asst	1887-1889	Wislicenus, W.F.	later lecturer, Prof Strasbourg
Asst	1889-1892	Zwink, Max	later at Imp. Patent Office
Asst	1892-1896	Wanach, B.	later at Potsdam Geodet. Inst.
Asst	1897-1898	Necker, K.	
Asst	1898-1901	Tetens, O.	later Samoa, Kiel and Berlin
Asst	1902	Ebert, W.	
Asst	1903	Carnera, L.	later Carloforte, Genoa and Trieste
Asst	1904-1910	Jost, E.	later Koenigsberg Obs.
2-Asst	1887-1888	Kaufmann, August	later secondary school teacher
2-Asst	1889-1892	Halm, J.	
2-Asst	1892	Martin, P.	
2-Asst	1893-1895	Halm, J.	later Asst Edinburgh
2-Asst	1895-1897	Necker, K.	
2-Asst	1897-1901	Ebell, M.	later at edit. off. *Astr. Nachr.*, Kiel
2-Asst	1902	Meyer, H.	
2-Asst	1903-1904	Cohn, B.	
2-Asst	1903-1904	Schiller, K.	
2-Asst	1904	Redlich, E.	later Asst Kiel
2-Asst	1905-1906	Biske, F.	later Warsaw Univ.
2-Asst	1907-1913	Liebmann, J.	
Asst-W	1904-1919	Cohn, B.	
Porter	1887-1904	Sabel, Carl	
Porter	1905-1919	Sause, Ewald	
Servant	1887-1888	Oelke, H.	
Servant	1888-1890	Kipp, J.	
Servant	1891-1892	Bauer, Markwart	
Servant	1893-1894	Schwenn, Ludwig	
Servant	1895-1899	Rolf, Julius	
Servant	1900-1901	Jenne, Adolf	
Servant	1901-1918	Libertus, Hubert	

Abbreviations: Dir = Director, Prof = Professor, Obs = Observer, Asst = Assistant, 2-Asst = Auxiliary or Second Assistant, Asst W = Assistant Worker [wissenschaftl. Hilfsarbeiter], Servant = Servant and Mechanic, Calc = Calculator, Porter = Porter and Calculator

Additional information: Bernhard Wanach, see: Mahnkopf (1929), p. 31-36; Otto Tetens moved to the observing station of the *Göttinger Gesellschaft der Wissenschaften* in Samoa in 1902; Luigi Carnera arrived in 1902 from the Astrophysical Observatory in Heidelberg; he went in 1904 to the Carloforte International Latitude Station, from 1910 to 1918 he was Professor at the Hydrographical Institute in Genoa, from 1919 Director of Trieste Observatory.

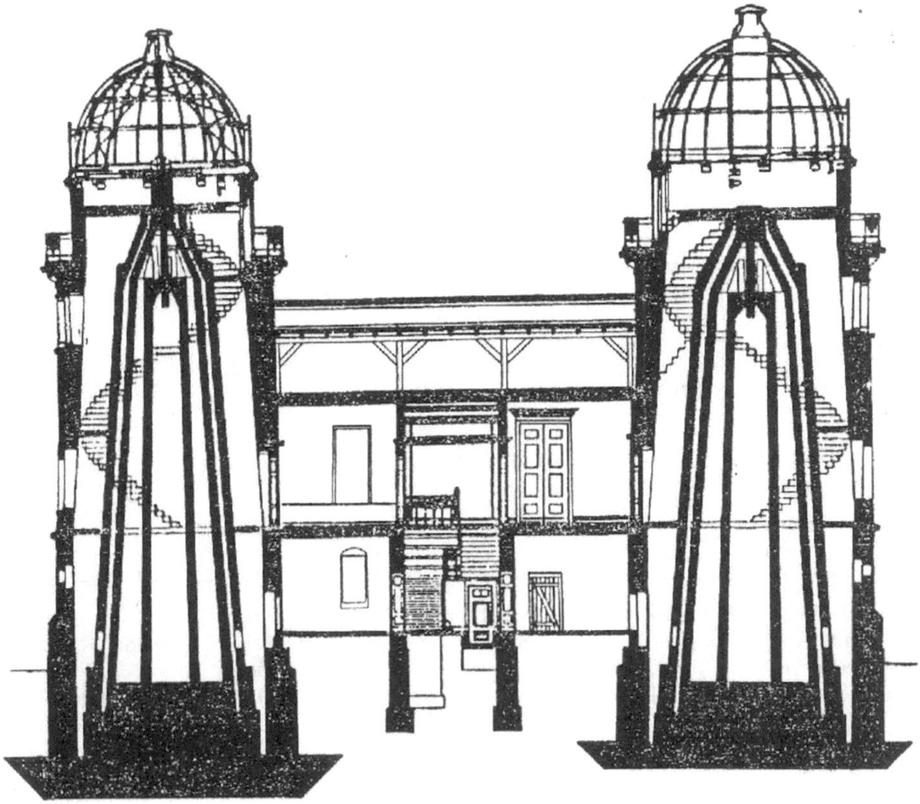

Figure 3. Observatory building with meridian room from North to South. (Müller 1978, p. 125)

teachers. In 1880 he moved to Strasbourg, where he stayed for nearly 30 years, first from 1883 to 1889 as an assistant of the observatory. After his habilitation in 1887 he became adjunct professor in Strasbourg in 1894. Starting in 1899, he edited the *Astronomischer Jahresbericht*, a bibliography of astronomical literature.

An attempt by the Strasbourg faculty of mathematics and science to create an adjunct professorship for astrophysics was rejected in 1895.

We cannot discuss the assistants in detail. The early staff of Strasbourg observatory, in the days of Winnecke and Becker, all succeeded in brillant careers using Strasbourg as a launching pad (cf. Tables 1 and 2).

Julius Bauschinger (1860-1934), born in Fürth in Bavaria, studied in Munich, where his father was professor at the Polytechnikum, and in Berlin. In 1883, after participation in a German Venus transit expedition, he obtained a position as assistant with Seeliger in Munich. As observer since

TABLE 3. Staff in Strasbourg – Bauschinger Era

Position	Year	Name	Notes
Dir	1909-1919	Bauschinger, J.	later Dir Leipzig
Obs	1902-1919	Wirtz, C.W.	later Obs Kiel
Asst	1904-1910	Jost, E.	later Obs Koenigsberg
Asst	1910-1919	Hassenstein, W.	later Asst, Obs Potsdam
Asst	1910-1913	Liebmann, J.	later Obs Babelsberg
Asst	1914-1919	Schiller, K.	later Asst, Obs Leipzig
2-Asst	1907-1910	Liebmann, J.	
Asst W	1904-1919	Cohn, B.	later Calc Strasbourg Obs
Asst W	1913-1914	Pummerer, P.	later Hamburg Seewarte
Calc	1915-1918	Kipp, A. (Miss)	
Calc	1915-1918	Krusch (Mrs.)	
Calc	1916-1917	Hamm, E.	
Servant	1901-1919	Libertus, Hubert	
Doorman	1905-1919	Sause, Ewald	

Abbreviations: Dir = Director, Obs = Observer, Asst = Assistant, 2-Asst = Auxiliary Assistant, Asst W = Assistant Worker [wissenschaftl. Hilfsarbeiter], Servant = Servant and Mechanic, Calc = Calculator, Porter = Porter and Calculator
Additional information: Hassenstein, Walter Georg: came 1910 from Koenigsberg, East Prussia, went in 1920 as assistant to the Astrophysikalisches Observatorium Potsdam, in 1922 he became observer; Schiller, Karl: 1907/13 director of Bothkamp observatory near Kiel, in 1920 assistant and in 1923 observer in Leipzig; 1946-1962 director in Frankfurt/Main.

1886 he was active in meridian circle astronomy (habilitation in 1888). In 1896, at the age of 36, he received the prestigious offer to be successor of Tietjen as the director of the *Astronomisches Recheninstitut* in Berlin, and also to hold a chair in theoretical astronomy, *i.e.* celestial mechanics. In 1906 his famous book *Bahnbestimmung der Himmelskörper* appeared, known in Germany and abroad as "the Bauschinger"; it became a standard work (2nd edition 1928). The following reasons – apart from some difficulties in Berlin – may explain his acceptance of the position of director in Strasbourg in 1909 (Stracke 1934):

> *He was tempted by the possibility to observe at Strasbourg Observatory, which was well-equipped in those times. He was tempted by southern Germany, with the Black Forest and the Vosges Mountains nearby, the less formal life in a middle-sized town, which offered to him and his family in an own house in the middle of a huge observatory garden more freedom of movement.*

In January 1919, Bauschinger had to leave Strasbourg; he moved to Munich, and in 1920 he succeeded Bruns in Leipzig, where he was director until 1930.

Bauschinger's staff (cf. Table 3) included Walter Georg Hassenstein, J. Liebmann and Karl Schiller[22] (1882-1979); they were active in the field of meridian circle astronomy; on the other hand the observer Carl Wilhelm Wirtz[23] (1876-1939) observed nebulae and binaries with the large refractor.

A quick look at the students and guests in Strasbourg, who were quite international (cf. Table 4), is also very interesting. But only students who were active in the institute, *i.e.* those who wrote their theses in Strasbourg, are mentioned here. All students, where no information is given, presumably left astronomy, *i.e.* they either moved to another branch of science such geodesy, geophysics or meteorology, or they became teachers in secondary schools.

5. Research and Teaching in Strasbourg in the German period

The unique organisational conditions in Strasbourg promoted team work, with the faculty of medicine as well as inside the faculty of science, crossing borders of institutes; even assistants were exchanged. Teamwork was normal, contrary to other universities. Winnecke tested and checked his objective for the large refractor using a spectrometer in the physical laboratory. Students listened to lectures of other faculties.

Die Fakultät muß aber Gewicht darauf legen, daß regelmäßige Vorlesungen über praktische Astronomie, sowie auch besonders über die theoretischen Theile dieser Wissenschaft stattfinden. es ist dies erforderlich nicht nur im Interesse derjenigen Studierenden, welche sich speziell der Astronomie widmen, sondern auch der sämmtlichen Studierenden der Mathematik.[24]

[The faculty has to consider it to be important that regular classes are offered both in practical astronomy, as well as in the theoretical fields of this science. This is necessary both for those students whose studies are particularly dedicated to astronomy, as well as for all students of mathematics.]

Founded as a reform university, it offered unique room for development for comparatively young professors and for new topics. Already during the search for the first director the authorities tried to hire Zöllner, as a pioneer in the new field of astrophysics. In 1895 the Faculty of Mathematics and Science asked again Strasbourg University to introduce astrophysics and to create at least an assistant "extraordinary" professorship (a.o. Professor) for astrophysics:

Als die hiesige Hochschule gegründet wurde, konnte es ausreichend erscheinen, für den astronomischen Unterricht einen Lehrstuhl zu errichten, dem die Aufgabe zufiel, die Hauptfächer in geeigneter Folge vorzutragen, und der freiwilli-

[22]Anonymous: Nachruf – Professor Dr. Karl Schiller (1882-1979). In: Jahresbericht des Physikalischen Vereins Frankfurt 154 (1981), p. 14-15.

[23]Duerbeck & Seitter (1990).

[24]Eingabe der mathematisch-naturwissenschaftlichen Fakultät ans Ministerium vom 8. März 1886. Arch. Dép. Bas-Rhin, AL 103, pag. 71, No. 245, Bl. 34.

TABLE 4. Students and Guests in Strasbourg 1872-1918

Year	Name/provenience (later postion)
1876-1879	Küstner, K.F. (Obs Hamburg, Obs Berlin, Prof, Dir Bonn)
1876-1877	Fuchs
1877-1878	Mahn
1877-1880	Elkin, W./New Orleans,LA,USA (1884 Cape, 1896 Dir Yale, USA)
1878	Branting/Stockholm
1878-1879	Bröker
1879	Rausenberger & Büngner (cand.)/Dorpat
1879	Struve, H./Pulkowa (1895 Prof, Dir Königsberg, 1904 Dir Berlin)
1878-1880	Ambronn, L. (1880 Seewarte Hamburg, 1897 Prof Göttingen)
1880	Isenbeck
1880-1883	Wislicenus, W.F. (1883 Asst Straßburg, 1894 Prof Straßburg)
1887-1889	Ristenpart, F.W. (1900 Akademie Berlin, Dir Santiago, Chile)
1887-1889	Schröter, J.F./Christiana (1922 Prof, 1926 Dir Oslo)
1888	Stutz & Reiss (cand.)
1888	Wolf, M. (1893 Dir Astrophys. Obs. Heidelberg, 1902 Prof)
1889-1890	von Burgesdorff
1890	Frost, E.B./USA (1895 Prof Dartmouth, 1898 Chicago, 1905 Yerkes)
1890-1891	Riem, J. (1905 Obs, 1913 Prof Berlin, Astr. Recheninstitut)
1890	Paetsch, H./Berlin (Prof Berlin)
1891	von Harkányi, Baron B. (1899 O'Gyalla, 1907 PD Budapest)
1891-1893	Schwarzschild, K. (1899 PD Munich, 1901 Prof, Dir Göttingen)
1892-1893	Kohlschütter, E. (1922 Prof Berlin, Dir Geodet. Inst. Potsdam)
1893-1896	Courvoisier, L. (1905 Obs Berlin-Babelsberg, 1926 Prof Berlin)
1894-1895	Peck, H.A./Mexico, N.Y., USA (1894 Prof Syracuse Uni)
1895	Breithaupt, Reiff, Rödiger, Mönichs & Ehlert
1902	Meyer, H.
1908	Plate, H. (cand.) & Jonckheere, R./Roubaix
1908-1914	Heinrich, W./Przibam (1919 a.o.Prof, Dir Prag, 1926 Prof Prag)
1910-1911	Tschetschott, A.
1910-1912	Benes, L./Prag
1915	Pavel, F. (cand.) (1926 Asst Berlin-Babelsberg)

gen Mitwirkung von jüngeren Kräften es zu überlassen, für gewisse spezielle Aufgaben ergänzend einzutreten. Inzwischen hat im Laufe der letzten beiden Jahrzehnte der Umfang der astronomischen Wissenschaft bedeutend zugenommen, neue Forschungsmittel und neue Wege sind gefunden, die es ermöglicht haben, Gebiete in Angriff zu nehmen, die bislang ganz verschlossen schienen.

Neben der Astronomie im üblichen Sinn ist die Astrophysik entstanden, die auf die Analyse des Lichtes gegründet, ihre Schlüsse aus der Spektrometrie und Photometrie ableitet. Anfänglich auf das weite, aber begrenzte Gebiet der Erforschung der chemischen und physischen Constitution der Himmelskörper beschränkt, hat sie im letzten Jahrzehnt, dank der weiteren Entwicklung der Theorie und begünstigt von den Fortschritten der photographischen Technik, auch für mannigfaltige allgemeine Aufgaben der Himmelsforschung äußerst fruchtbringend sich erwiesen, die glänzenden Erfolge, die sie schon, selbst in der vornehmsten und schwierigsten Aufgabe der Astronomie, der Erkenntniß der Anordnung des Weltsystems errungen hat, berechtigen zu neuen großen Erwartungen.

Bei dieser täglich wachsenden Bedeutung der Astrophysik erscheint es unumgänglich, auch im Unterricht eine gebührende Stelle ihr [einzuräumen?] und vor allem die Studierenden in die grundlegenden Lehren dieser Disciplin gehörig einzuführen. Es darf und kann dies aber nicht unter Einschränkung des bisherigen astronomischen Unterrichts geschehen; denn nicht nur bedarf auch der Astrophysiker der eingehenden Kenntnis der älteren Methoden und wird durch die Gewöhnung an ihre Exactheit von der nahe liegenden Gefahr leichtfertiger Schlußfolgerungen und vager [Schätzungen?] bewahrt, sondern sie behalten auch neben den neuen Methoden ihre selbständige Bedeutung und erfordern eine weitere Ausbildung, wenn die Ergebnisse der letzteren voll und ganz verwertet werden sollen.

Überzeugt, daß es dringend wünschenswert ist, daß der Astrophysik im Unterrichtswesen unserer Hochschule daher eine besondere und angemessene Stelle eingeräumt werde, gestattet sich der Unterzeichnende der Fakultät die Einrichtung einer a.o. Professur zu empfehlen.[25]

[When this university was founded, it may have appeared sufficient to install a professorial chair dedicated to the task to present the main topics in suitable succession, and to leave it to the younger lecturers to volunteer in supplementing them by special tasks. In the course of the last two decades, the range of astronomical science has widened considerably, new means of research and new pathways have been discovered that have made it possible to attack problems that seemed to be completely untackable.

Besides astronomy proper, astrophysics has emerged, which is based on the analysis of light, and which draws its conclusions from spectroscopy and photometry. In the beginning, it was restricted to the wide but limited field of the chemical and physical constitution of celestial bodies, but in the last decade, thanks to the progress in photographic techniques, it has proven extremely fertile for manifold general problems of celestial research. The splendid success which it has achieved in the most prominent and most difficult astronomical task, the knowledge if the constitution of the sidereal system, permits to expect new great discoveries.

In view of this continuously growing importance of astrophysics it appears absolutely necessary to give it a prominent place also in teaching, and to properly introduce especially the students in the basics of this discipline. This should not be done by restricting the present astronomical instruction; since not only the astrophysicist needs an intense knowledge of the older methods, and is, by getting accustomed to their exactness, saved from the peril of wanton conclusions and vage estimates, they also keep their own importance besides the new methods and require a deeper study, if the results of the latter should have their most fruitful value.

[25]Schreiben der Fakultät an den Statthalter, 27.6.1895. Arch. Dép. Bas-Rhin, AL 103, pag. 71, No. 245, Bl. 195-196.

The undersigned is pleased to suggest to the faculty the installment of an extraordinary professorship, since he is persuaded that it is urgently desirable to give astrophysics in the instrucion plans of our university a special and appropriate place.]

This application of the faculty was submitted via the dean and the curator to the *Statthalter* in June 1895. But it was declined with the explanation that in Germany astrophysics existed only in one university with one assistant professor – Julius Scheiner (1858-1913) in Berlin –, that in Göttingen only the private lecturer gave lectures in this field and that in Leipzig the chair of Zöllner has not been renewed, "sie scheine eingegangen zu sein"[26] [it seems to have vanished]. Thus for the persons who were interested in modern astrophysics had only the possibility to listen to the lectures of the private lecturer Franz Stenger (1859-1893) about spectral analysis in physics which were offered in the winter term 1886/87.

The emphasis in research and teaching was in all fields of classical astronomy, which can be seen quite well by looking at the lectures of Winnecke and Schur (cf. Table 5 and 6). Winnecke was mainly engaged with the determination of positions of nebulae and comets as well as with variable stars – mostly by using Argelander's step estimate method for visual photometric measurements, not even with the help of an instrument such as Zöllner's photometer.

Schur observed the moons of Jupiter and double stars with the heliometer and he made timekeeping measurements with a transit instrument. Hartwig had two work areas: on the one hand he made astrometric observations with the heliometer with which he also determined parallaxes of fixed stars as well as diameters of planets and the physical libration of the moon. On the other hand he discovered two comets (1879 IV and 1888 III) and in 1885, after his departure to Dorpat, a 'new star' (S Andromedae), which later turned out to be the first extragalactic supernova.

Also the lectures of Ernst Becker and private lecturer Kobold show the broad spectrum of classical astronomy (cf. Table 7 & 8); in the case of Becker there is an emphasis on orbit determination and calculation of perturbations. Participation in the German Venus transit expedition in 1874 led Becker to an additional field of interest, the exact determination of the diameter of the Sun with a heliometer (and an examination if there are changes with time). Kobold investigated comets, planetary moons and nebulae with the large refractor.

[26]The correspondence is given in: Arch. Dép. Bas-Rhin, AL 103, pag. 71, No. 245, Bl. 189 Dekan an Kurator mit der Bitte um Weiterleitung des Gesuchs an den Statthalter 28.6.1895, Bl. 189r Kurator an Dekan 11.8.1895, Bl. 194 Dekan an Kurator 12.9.1895, Bl. 190/191 Kurator an Statthalter 17.9.1895, Bl. 192 Statthalter an Kurator 26.9.1895, Bl. 192r/193 Kurator an Fakultät 1.10.1895.

TABLE 5. Winnecke's lectures at Strasbourg University

Topic	Term (Semester)
Elemente der (praktischen) Astronomie	WS 73/74, WS 75/76, SS 77, WS 77/78
Einleitung in die (praktische) Astronomie	WS 79/80, SS 81, WS 81/82
Astrognostische Übungen/Astrognosie	SS 73, SS 76
Sphärische Astronomie	SS 73, SS 74, SS 76, SS 78
Theoretische Astronomie	WS 74/75, WS 76/77, WS 78/79, WS 80/81
Astronomische Methoden zur geographischen Ortsbestimmung	SS 75
Praktische Übungen an den Instrumenten der Sternwarte	WS 73/74, WS 74/75, WS 75/76, WS 76/77, SS 77, SS 78, WS 78/79, WS 79/80, SS 80, WS 80/81, SS 81, WS 81/82, SS 82
Praktische Übungen an den Instrumenten der Sternwarte für Geübtere	SS 75, SS 76, WS 77/78
Topographie des Sonnensystems	SS 74, WS 76/77, SS 78, WS 79/80, WS 81/82
Kometen(kunde)	SS 75, SS 77, SS 82
Theorie der Kometenschweife	WS 75/76
Fixsternkunde	WS 74/75, WS 77/78, WS 80/81
Theorie der Venus-Vorübergänge	SS 77, SS 80, SS 82

Notes: The terms SS 72 and SS 79 are missing. In SS 74 there was offered in addition: Übungen mit Anschluß an die Vorlesung Topographie des Sonnensystems. SS 77 and SS 80: Theorie der Vorübergänge der unteren Planeten vor der Sonne mit besonderer Rücksicht auf die Venusdurchgänge von 1874 und 1882. SS 82: Vorübergang der Venus 6. Dec. 1882.

The only exception was the teaching of the private lecturer Wislicenus, who, starting in 1889, offered in addition to classical astronomy modern topics in astrophysics, like spectroscopy and photometry (cf. Table 9). In addition, in the winter term 1890/91, a lecture was given by the private lecturer in physics Otto Wiener (1862-1927) about spectral analysis. Wislicenus started with photometric work concerning the Moon and comets, which was continued by Wirtz concerning nebulae; in addition he used an actinometer for the measurement of solar radiation. Karl Schiller was a specialist in variable stars.

Besides his university lectures, Wislicenus also gave popular talks to the general public (Kobold 1906, p. 20):

> *This rich literary activity was supplemented by his lectures, which were much appreciated and well attended, and which took place not only in*

TABLE 6. Schur's lectures at Strasbourg University

Topic	Term (Semester)
Geographische Ortsbestimmungen auf Reisen	SS 82
Bestimmung von Doppelstern-Bahnen	WS 82/83
Methode der kleinsten Quadrate	WS 82/83, SS 84, SS 85
Theorie der Berechnung planetarischer Störungen	WS 83/84
Praktische Übungen an den Instrumenten der Sternwarte	WS 83/84, SS 84, WS 84/85, SS 85
Praktische Astronomie	WS 84/85

Notes: The terms SS 83, WS 85/86 are missing. Further lectures: Theorie und Geschichte der Bestimmung von Doppelstern-Bahnen, WS 82/83. Praktische Anwendung der Methode der kleinsten Quadrate auf die Berechnung von Beobachtungsresultaten, WS 82/83. Methode der kleinsten Quadrate mit Anwendungen, SS 84. Sphärische Astronomie mit Übungen im Gebrauch der astronomischen Instrumente und in der Berechnung astronomischer Beobachtungen, SS 88. Sphärische Astronomie mit besonderer Berücksichtigung der mikrometrischen Messungen, SS 89 and SS 91.

TABLE 7. Becker's lectures at Strasbourg University

Topic	Term (Semester)
Allgemeine Astronomie	WS 88/89, WS 91/92
Sphärische Astronomie (mikrometr. Messungen)	SS 88, SS 89, SS 91
Sphärische und Praktische Astronomie	SS 90
Theoretische Astronomie	WS 90/91
Praktische Übungen auf der Sternwarte	WS 88/89, SS 89, WS 89/90, SS 90, SS 91
Übungen in astronomischen Beobachtungen	WS 90/91, WS 91/92
Methode der kleinsten Quadrate	SS 88
Ableitung der Elemente von Planetenbahnen	WS 88/89
Specielle Störungen und Bahnverbesserung	WS 91/92
Theorie der Störungen (Methode von Hansen)	WS 91/92
Über Sternkataloge (und ihren Gebrauch)	SS 89, SS 91
Über Doppelsterne und Berechnung ihrer Bahnen	WS 89/90
Mechanik des Himmels in elementarer Darstellung	WS 89/90
Theorie des Heliometers	SS 90
Anwendung der Wahrscheinlichkeitsrechnung auf Beobachtungen und Messungen	WS 90/91

Further topics: Einleitung in die Sphärische und Praktische Astronomie. Methode der kleinsten Quadrate und ihre Anwendung auf astronomische Rechnungen. Specielle Störungen und Ableitung der wahrscheinlichen Elemente einer Planeten- oder Kometenbahn. Einleitung in die Theorie der allgemeinen Störungen nach der Methode von Hansen.

TABLE 8. Kobold's lectures at Strasbourg University

Topic	Term (Semester)
Über die Methoden der Parallaxenbestimmung	WS 88/89, SS 89, WS 91/92
Astronomische Geographie	WS 89/90
Geodäsie	SS 90
Theorie der Finsternisse	WS 90/91
Theorie der Planetenvorübergänge	SS 91

TABLE 9. Wislicenus' lectures at Strasbourg University

Topic	Term (Semester)
Geographische Ortsbestimmungen auf Reisen	SS 88, SS 90
Geschichtliche Einleitung in die Astronomie	WS 88/89, SS 90
Geometrische Optik	WS 88/89, WS 90/91
Dioptrik (Fernrohr und Mikroskop)	SS 89, SS 91
Photometrie des Himmels	SS 89, WS 89/90, WS 91/92
Kalendarische und chronologische Rechnungen	WS 89/90, WS 91/92
Astrophysik des Sonnensystems	WS 90/91
Spectroscopie des Himmels	SS 91

Further Topics: Anweisung zu den einfachsten kalendarischen und chronologischen Rechnungen.

his narrower circle of activity in Strasbourg, but also in other towns. They became a rich source of instruction for wider circles about the more recent results of astronomical research, since he put much effort and love in preparing them.

To describe the activities in the Strasbourg observatory of that time we can give the following quotation, which refers to Bauschinger but characterises all of the Strasbourg research very well (Hopmann 1934, p. 299):

Bauschinger was the representative of "classical astronomy" with a purity that rarely exists, that field of research which limits itself to the determination of stellar positions, motions, and masses, which only applies Newtonian mechanics and its consequences, and completely leaves aside the physical possibilities of research that are available today. On the solid foundations of Newton the 18th century theoreticians, and then especially Gauss and Bessel, have brought spherical astronomy, orbit determinations of planets and comets, and celestial mechanics to mathematical and computational perfection, so that major challenges were ba-

sically lacking since the middle of the 19th century. It was Bauschinger's aim to refine all this by means of observations, by improving the mathematical formalism, and by organizing scientific work, and it was crowned by success.

6. The World War II Period

At the outbreak of World War II, André Danjon (1890-1967) was vice-president of the university. In November 1939 he organized the necessary relocation of the university to Clermont-Ferrand (including the Observatory):[27]

> Because of the evacuation of Strasbourg in 1939, the meridian circle, the main clocks and the optics of the major instruments had been brought to Paris and Bordeaux, respectively, the library and the major part of the workshop, as well as the smaller instruments and auxiliary apparatus to Clermont-Ferrand[28].

During World War II, when Johannes Hellerich had taken over the directorship of the Strasbourg observatory, the research after the recirculation of the instruments was practically not started, but the time passed by, like it could be read in the annual reports, with the re-establishment of the observatory and with the restoration and installation of the instruments.

Johannes Hellerich[29] (1888-1963), born in Hamburg, studied in Jena and Berlin and was awarded his PhD in 1913 with a thesis about variable stars. He was hired as assistant in Bonn. After World War I, he became an assistant in Kiel in 1919 where he wrote his habilitation about eclipsing binaries in 1924. In 1928, he was named as professor. In 1929, he became scientific observer ('Observator') in Hamburg-Bergedorf. In 1941, he became 'UK' ('unabkömmlich', in a reserved occupation) and was appointed as full professor for astronomy and astrophysics and director of Strasbourg Observatory.

The small staff of Hellerich in 1941 included the assistant J. Huß (aide-astronome under Danjon!), the (female) computer and librarian E. Rühl, the precision mechanic R. Schneider, the technical employee A. Ehrhart and the laboratory assistant, gardener and concierge L. Wenzel.[30] In 1942 for a short time, the scientific assistant K. Schrey was employed, until he left for military duty and was replaced by Miss A. Breitung.[31] After the end of the war and after a short internment, Hellerich was dismissed in

[27]Fehrenbach (1990), p. 223-245.
[28]Hellerich (1942), p. 215-216.
[29]Straßl (1964), p. 27-34.
[30]Hellerich (1942), p. 215-216.
[31]Hellerich (1943), p. 181.

1946 and went to Hamburg[32]. This short second German period stopped in November 1944.

7. Conclusion

As a conclusion one can say that Strasbourg observatory attained international status as a model observatory during its first German period and offered unique working conditions.

Acknowledgement

I am grateful to James Caplan for his help in translating this article from German into English, as well to the Editor, to an anonymous referee and to H.W. Duerbeck for their suggestions. I am also indebted to the latter for the translations of the German quotations into English.

Archive Material

Archives Départementales du Bas-Rhin, Strasbourg (Archives pour la Région Alsace):

Eingabe der mathematisch-naturwissenschaftlichen Fakultät ans Ministerium vom 8. März 1886, AL 103, pag. 71, No. 245, Bl. 34

Berufungsvorgang erster Director AL 103, pag. 71, No. 245, Bl. 34 to 41, p. 42 to 71, Bl. 72 bto 76.

Seeliger, Hugo von: Schreiben an die Universität Straßburg, Unterstaatssekretär Ledderhose, 13.6.1886, AL 103, pag. 71, No. 245, Bl. 40.

Eingabe der mathematisch-naturwissenschaftlichen Fakultät ans Ministerium vom 8. März 1886 , AL 103, pag. 71, No. 245, Bl. 34.

Schreiben der Fakultät an den Statthalter, 27.6.1895, AL 103, pag. 71, No. 245, Bl. 195-196.

Dekan an Kurator mit der Bitte um Weiterleitung des Gesuchs an den Statthalter 28.6.1895, AL 103, pag. 71, No. 245, Bl. 189

Kurator an Dekan 11.8.1895, AL 103, pag. 71, No. 245, Bl. 189, Bl. 189r

Dekan an Kurator 12.9.1895, AL 103, pag. 71, No. 245, Bl. 189, Bl. 194

Kurator an Statthalter 17.9.1895, AL 103, pag. 71, No. 245, Bl. 189, Bl. 190/191

Statthalter an Kurator 26.9.1895, AL 103, pag. 71, No. 245, Bl. 189, Bl. 192

Kurator an Fakultät 1.10.1895, AL 103, pag. 71, No. 245, Bl. 189, Bl. 192r/193

[32]In Hamburg, he became a scientific assistant in 1946. In 1949, he became assistant professor ('Extraordinarius') and director of the Astronomical Institute of Münster University. The instruments, dome and library in Münster had been destroyed during the war, so Hellerich, together with an assistant, used temporary working rooms; in 1956, he became 'Emeritus' (he has retired).

Universitätsarchiv Leipzig:

Zöllner, Karl Friedrich: Personalakte.

References

1. AIV 1894, Straßburg und seine Bauten, Architekten- und Ingenieurverein für Elsaß-Lothringen, Straßburg.
2. Auwers, A. 1881, Bericht über die Versammlung der Astronomischen Gesellschaft zu Strassburg 1881 September 22 bis 24, *Vierteljahrsschrift Astron. Ges.* **16**, 261-289.
3. Becker, E. (Ed.) 1896, Annalen der kaiserlichen Universitätssternwarte in Straßburg, Band 1, Karlsruhe.
4. Bodenmuller, W. & Heck, A. 2005, Strasbourg Observatory, Astronomical Phenomena and the Regional Press, this volume.
5. Brendel, M. 1902, Adolph Christian Wilhelm Schur, *Vierteljahrsschrift Astron. Ges.* **36**, 164-170.
6. Duerbeck, H.W. & Seitter, W.C. 1990, Carl Wirtz, die extragalaktischen Nebel und die frühe relativistische Kosmologie I-III, *Die Sterne* **66**, 3-15, 81-94, 131-139.
7. Ebell, M. 1942, Hermann Kobold, *Vierteljahrsschrift Astron. Ges.* **77**, 241-252.
8. Eggert, H. 1884, Festschrift zur Einweihung der Neubauten der Kaiser-Wilhelms-Universität Straßburg, Strasbourg.
9. Eggert, H. 1887, Die Sternwarte in Bamberg, *Zentralblatt der Bauverwaltung* **7**, 353-354.
10. Fehrenbach, Ch. 1990, Des hommes, des téléscopes, des étoiles, Ed. CNRS, Paris.
11. Foerster, W. 1898, August Winnecke, *Himmel und Erde* **10**, 230-236.
12. Hartwig, E. 1887, Ueber die Bamberger Sternwarte, *Vierteljahrsschrift Astron. Ges.* **22**, 277 & 329-335.
13. Hartwig, E. 1898, Friedrich August Theodor Winnecke, *Vierteljahrsschrift Astron. Ges.* **33**, 5-13.
14. Heise, E. 1923, Hartwig, Ernst, *Deutsches Bibliographisches Jahrbuch*, 154-157.
15. Hellerich, J. 1942, Jahresbericht der Sternwarte Straßburg für 1941, *Vierteljahrsschrift Astron. Ges.* **77**, 215-216.
16. Hellerich, J. 1943, Jahresbericht der Sternwarte Straßburg für 1942, *Vierteljahrsschrift Astron. Ges.* **78**, 181.
17. Hoffmeister, C. 1923, Ernst Hartwig, *Astron. Nachr.* **219**, 185.
18. Hopmann, J. 1934, Nachruf auf Julius Bauschinger, *Berichte und Verhandlungen der Sächsischen Akademie der Wissenschaften, math.-phys. Klasse* **86**, Nr. IV, 299-306.
19. Jost, E. 1913, Ernst Becker, *Vierteljahrsschrift Astron. Ges.* **48**, 2-12.
20. Kobold, H. 1906, Walter Wislicenus, *Vierteljahrsschrift Astron. Ges.* **41**, 13-21.
21. Koerber, F. 1899, Karl Friedrich Zöllner. Ein deutsches Gelehrtenleben, Berlin (Sammlung populärer Schriften herausgegeben von der Gesellschaft Urania zu Berlin).
22. Mahnkopf, H. 1929, Bernhard Karl Wanach, *Vierteljahrsschrift Astron. Ges.* **64**, 31-36.
23. Mayer, O. 1922, Die Kaiser-Wilhelms-Universität Straßburg. Ihre Entstehung und Entwicklung, Vereinigung wissenschaftlicher Verleger, Berlin, Leipzig.
24. Meinel, Chr. 1991, Karl Friedrich Zöllner und die Wissenschaftskultur der Gründerzeit. Eine Fallstudie zur Genese konservativer Zivilisationskritik, Berlin: ERS-Verlag (Berliner Beiträge zur Geschichte der Naturwissenschaften und der Technik, 13).
25. Meyermann, B. 1902, Verzeichnis der Schriften von W. Schur, *Vierteljahrsschrift Astron. Ges.* **36**, 170-172.

26. Müller, P. 1975, Sternwarten. Architektur und Geschichte der Astronomischen Observatorien, Lang Verlag, Frankfurt (Europäische Hochschulschriften, Reihe XXXII: Astronomie, Bd. 1), (2. Auflage) 1978, 170-175.
27. Schur, W. 1886, Bericht über die an den Instrumenten der Strassburger Sternwarte in den letzten Jahren ausgeführten Untersuchungen und Beobachtungen, *Astron. Nachr.* **112**, 161-176.
28. Schur, W. 1886, Fernerer Bericht über die Thätigkeit der Strassburger Sternwarte, *Astron. Nachr.* **114**, 401-404.
29. Seggewiss, W. 2005, Strasbourg Observatory and the Astronomische Gesellschaft, this volume.
30. Spieker, P. 1879, Die Bauausführungen des Königlichen Astrophysikalischen Observatoriums auf dem Telegraphenberge bei Potsdam, *Zeitschrift für Bauwesen.*
31. Stracke, G. 1934, Julius Bauschinger, *Vierteljahrsschrift Astron. Ges.* **69**, 146-163.
32. Straßl, H. 1964, Johannes Hellerich, *Mitteilungen Astron. Ges. für 1963*, 27-34.
33. Struve, F.G.W. 1845, Description de l'observatoire astronomique central de Poulkova, Petersburg.
34. Zöllner, K.F. 1865, Photometrische Untersuchungen mit besonderer Rücksicht auf die physische Beschaffenheit der Himmelskörper, Engelmann, Leipzig.
35. Zöllner, K.F. 1881, Wissenschaftliche Abhandlungen mit zwei populären Vorträgen über die universelle Bedeutung der mechanischen Principien und die physische Beschaffenheit der Sonne, **Band 4**, Staackmann, Leipzig.

THE OBSERVATORY OF THE EMPEROR WILHELM UNIVERSITY: THE PEOPLE BEHIND THE DOCUMENTS

H.W. DUERBECK
Vrije Universiteit Brussel
Pleinlaan 2
B-1050 Brussel, Belgium
hduerbec@vub.ac.be

Abstract. We give an overview of the personnel of Strasbourg Observatory between 1872 and 1919, by making use of official documents, recollections of Wilhelm Foerster and Hermann Kobold, reports and bibliographies. The careers of almost all scientific workers are briefly described, and supplemented by illustrations. In addition, the major projects carried out at the observatory are outlined, including those which were continued after the first world war, and the experiments which led to the development of a seismological station in Strasbourg. An Appendix includes the bibliography of PhD theses and of annual reports.

1. Winnecke and the Foundation of Strasbourg Observatory

After the end of the Franco-German war of 1870/71, France had to cede the borderlands Alsace and Lorraine to Germany. After a short time of occupation, when the region was administered by a military regime ["Generalgouvernement"], it was incorporated, in September 1871, as a special province ["Reichsland"] into the new German Empire. The region was administered from Berlin by an imperial authority [Reichsamt für Elsaß und Lothringen].

Strasbourg University was called into existence through a founding document of the German Emperor Wilhelm I, dated 28 April 1872. (In 1877, after the death of the emperor, it was renamed into Emperor Wilhelm University). It should become *a German university of the first rank, a respectable home for the German intellect and guardian of the German sci-*

A. Heck (ed.), The Multinational History of Strasbourg Astronomical Observatory, 89–122.

entific method[1]. In order to attract not only students from Alsace, but also from "old Germany" as well as from abroad, it should have a large faculty (Roggenbach's plans envisaged 124 professors "of distinction", at a time when Germany's largest universities, Berlin and Leipzig, had 102 professors each), an innovative curriculum, as well as support fellowships for advanced students. Not all of Roggenbach's suggestions were accepted, and he resigned from his post soon after the inauguration of the University in May 1872.

According to Roggenbach's plans, the university should have a chair of astronomy and an institute which complied to modern teaching and scientific research. *And in November [1871] someone persuaded Roggenbach to revise his proposals to include a chair of astrophysics, apparently Germany's first*[2]. Friedrich Zöllner of Leipzig University was the only "academic" astrophysicist in Germany, who was also known to have a difficult personality. Thus it was a somewhat daring task for the decision-making people in Berlin, but Zöllner declined when Leipzig endowed him with a professorship of astrophysics in early 1872. Thus a compromise was attempted in chosing an astronomer who would promise to bring a modern approach in research, although with somewhat classical methods.

Wilhelm Foerster (1911) writes in his memoirs: *The Imperial Government has acted directly with the new founding and equipment of Strasbourg University – the later State Secretary for Alsace-Lorraine, [Karl] Herzog, was active with special love and wisdom – as it was later with the founding of the Imperial Physico-technical Institute [Physikalisch-Technische Reichsanstalt].*

The equipment of Strasbourg University with well-endowed scientific institutes, especially also with a new observatory, also offered the possibility to bring into play the excellent astronomer August Winnecke, [...] both for the sake of his own work, as well as for the progress of science. The observatory which he was able to install with the means that were given to him became a special and permanent ornament of German science.

The first director of Strasbourg Observatory was thus to become Friedrich August Theodor Winnecke (Fig. 1). He was born on 5 February 1835 in Gross-Heere near Hildesheim, where his father was a protestant cleric. He studied astronomy in Goettingen (1853) and Berlin (1854), where he obtained his PhD on 7 August 1856 with a thesis on the double star η Coronae Borealis (Winnecke 1856). In Berlin, he became friends with the observa-

[1][Baron Franz von] Roggenbach [who was in charge of founding the new university] to the Imperial Chancellery, 30 September 1871 Arch. Dép. Bas-Rhin, AL 12, paq. 7; quoted by Craig (1984), p. 41.

[2]Roggenbach to the Imperial Chancellery, 30 November 1871; ibid., quoted by Craig (1984), p. 45.

Figure 1. August Winnecke (Spamer 1881).

tory assistant (and later director) Wilhelm Foerster. Winnecke spent from Autumn 1856 to Spring 1858 at Bonn Observatory, working under Argelander, where his interest in observing variable stars was spurred, and where he used the heliometer to measure parallaxes of the star Lalande 21185 and of the planetary nebula $h2241$, and to measure positions of Praesepe stars. A visit of Wilhelm Struve in Bonn, in Autumn 1857, led to an invitation to Pulkovo. From July 1858, Winnecke was adjoint astronomer at Pulkovo, and later became senior astronomer and vice-director (a sickness of the director Otto Struve put a heavy administrative load on him). He stayed there for six years, also marrying Hedwig Dell, a niece of Otto Struve.

Because he showed signs of a psychic illness *because of too intensive work and a certain hereditary disposition* (Foerster 1911), he had to quit work in late 1864. After several stays in various spas had proven unsuccessful, he asked for retirement at the end of 1865, returned to Germany and sought treatment in a hospital in Bonn-Endenich, run by Dr. Hertz.

Foerster (1911, p. 83) tells a touching story in his memoirs: *I especially recall a little scene which I experienced in the Bonn sanatorium, when I had introduced the sick friend. Mrs. Hertz presided at the lunch table, where most diverse patients in vivid conversation were grouped. She had placed the two newly-arrived guests, the sick and the sound astronomer, to her right- and left-hand side. After the guests had risen from the table, she told me with special delight that Professor Foerster, who had talked vividly, and Professor Winnecke, who had looked solemn and melancholic, had found a peculiar appreciation with the guests at the table: I had been deemed as the patient and the solemn friend as my guide and mentor.*

Within a year's time, in 1867, Winnecke's health was restored, and he moved with his family to Karlsruhe where, in the following 5 years, he was absorbed by private research. He also served as a member of the board of directors of the *Astronomische Gesellschaft*, and was among the founding members of the German commission for the preparation of the observation of the Venus transits 1874 and 1882, which envisaged to send out four or five expeditions to various corners of the earth to observe the transits.

In Karlsruhe, he was called on 13 September 1872 by State Secretary [Staatsminister] Karl Herzog, to accept the position of a full professor at Strasbourg University, and to develop plans for the construction of a new observatory. Since Herzog was also a good acquaintance of W. Foerster, it is very likely that Foerster was instrumental in suggesting Winnecke as a suitable candidate.

Here is the official letter, kept in the Prussian Secret State Archive, addressed to the Emperor and King, and dated Berlin, 9 September 1872:

Die Errichtung eines Lehrstuhls für Astronomie an der Universität Straßburg ist Bedürfniß. Für diesen Lehrstuhl scheint der als hervorragender Astronom bekannte Dr. Winnecke in Carlsruhe nach seinen bisherigen Leistungen besonders geeignet. Es sind daher mit demselben Verhandlungen geführt worden, in Folge derer er sich zur Uebernahme einer ordentlichen Professur für Astronomie in der philosophischen und naturwissenschaftlichen Fakultät der Universität bereit erklärt hat. [...]

Eure Majestät bitte ich hiernach allerunterthänigst, durch huldreiche Vollziehung der angeschlossenen Bestallungen den Dr. Winnecke zum ordentlichen Professor in der philosophischen und naturwissenschaftlichen Fakultät der Universität Straßburg Allergnädigst ernennen zu wollen[3].

[The installation of a professorial chair for astronomy at Strasbourg University is a necessity. It seems that Dr. Winnecke, in Karlsruhe, who is known as an eminent astronomer, is especially suitable according to his previous accomplishments. Therefore negotiations have been carried out with him, resulting in his declara-

[3]Geheimes Staatsarchiv PK, I Rep 89, No. 21693, p. 43.

tion of readiness to take over the post of university Professor of Astronomy in the Faculty of Philosopy and Natural Sciences at the University. [...]

I therefore ask your majesty most obediently, by benevolent signing of the enclosed appointment, to express your wish to most graciously appoint Dr. Winnecke as a university Professor of Astronomy in the Faculty of Philosopy and Natural Sciences at Strasbourg University.]

The appointment was approved by the Emperor on 13 September 1872. From 1874 onward, Winnecke started to work on his principal research project, his survey of nebulae (Winnecke 1875), at first with the 6-inch refractor at the old academy observatory[4].

Besides work on his nebular project, and the teaching of students, he was also instrumental in planning the calibration observations of the heliometers used for the Venus transit observations, and indeed a lot of such heliometer work was carried out in Strasbourg around 1874 and 1882; it was partly published in the reports of the German Venus Transit Commission (Auwers 1887-1898).

After acceptance of his suggestions for a new observatory, and the granting of the necessary means by the government, plans were made by the governmental architect [Regierungsbaumeister] Hermann Eggert. It was decided to build in the vicinity of the main university building a natural science campus in an area which had become available in the course of the extension of the city. This area, called *Am Fischerthor*, comprised the buildings of the physical, chemical, pharmaceutical, and botanical institute (including a botanical garden), as well as the Observatory, and in addition, a building for the institutes of zoology, mineralogy, geography, paleontology, petrography, as well as the geological state institute.

Construction of the meridian building were started in 1877, the other two (the great refractor and the living quarters) in the two subsequent years. Construction was finished in 1880, and the main instruments were set up in Winter 1880/81. In the Summer term 1881, classes and lab courses were given in the new building for the first time, and during 22-24 September 1881, the *Astronomische Gesellschaft* held its annual meeting in the vestibule of the refractor building. Foerster wrote (1911, p. 175): *When attending the Strasbourg meeting, I felt especially pleased to see my friend Winnecke in a most successful activity, and to enjoy the great impression that the shining new observatory, donated by the German Empire to Winnecke, made to the astronomers of all countries, including those from France.*

A description of the observatory is given both in the first volume of the Annals of the Observatory, in a Festschrift on occasion of the 58th meeting of German scientists and doctors, in 1885 (Schur 1884), as well as in a

[4] An overview of Winnecke's nebular project is given in Duerbeck & Seitter (2005).

historical account by Hausmann (1897). Here, the cost of the observatory is also specified: main building (great refractor) 194,000 M, meridian building (with refractors and meridian instruments) 215,000 M, apartment house for employees and connecting buidings between houses, 105,000 M[5].

The instruments were the following: the great refractor (lens diameter 487mm, focal length 7m), was manufactured by Repsold, Hamburg, and its optics by Merz, Munich; the meridian circle (lens diameter 162mm, focal length 1.9m) was manufactured by Repsold, Hamburg. The passage instrument, manufactured by Cauchoix, Paris, was taken from the old academy observatory. The small refractor [Bahnsucher] (lens diameter 162mm, focal length 1.8 m) was manufactured by Repsold, Hamburg. The altazimuth (lens diameter 135mm, focal length 1.6m) was also manufactured by Repsold, Hamburg. A heliometer, built by Fraunhofer's workshop, was acquired in 1877 from Gotha Observatory; both the heliometer and a small passage instrument were housed in small buildings.

Three students obtained a PhD under Winnecke's guidance (Fig. 2); they all graduated in 1880 (see the Appendix for a list of all PhD theses in astronomy awarded by the Emperor Wilhelm University): William Elkin (1880), a future director of Yale Observatory in New Haven, Ernst Hartwig (1880), who became the first director of the Dr. Remeis Observatory, Bamberg, Franconia, and Friedrich Kuestner (1880), who would later become director of Bonn Observatory.

At the end of 1881, Winnecke was offered the post of director of Munich Observatory as Lamont's successor, but he declined. In February 1882, around the time of his election to the post of Rector of the University, Winnecke began again to suffer from his psychic illness. It was presumably triggered by the loss of his eldest son, who had drowned while ice-scating a few weeks before. Again, he turned to the sanatorium in Bonn, but this time no improvement was reached. In 1886, the vacancy at the head of the observatory became too obvious. Kobold (2004) writes in his autobiography: *...his state had deteriorated in such a way that he had to be declared incurable. He suffered from attacks of madness and had the illusion that his whole family was dead; when his wife and children visited him, they were only permitted to talk to him behind a lattice.* Winnecke was persuaded to ask for his dismissal, which was granted on 31 May 1886; only afterwards the search for a successor could be started. Winnecke died in Bonn, eleven years after his retirement, on 2 December 1897.

Here is a translation of the document, dated Strasbourg, 24 May 1886:

Der ordentliche Professor in der mathematischen und naturwissenschaftlichen Fakultät der Kaiser-Wilhelms-Universität Straßburg und Direktor der Universitäts-

[5]The total sum of 514,000 Marks corresponds to about 2 million Euros of today's purchasing power.

Figure 2. William Elkin, Ernst Hartwig and Friedrich Kuestner, Strasbourg's first three PhDs in astronomy (Shane Archives and PGAG 1904).

Sternwarte Dr. Winnecke ist zu Anfang des Jahres 1882 in eine Gemüthskrankheit verfallen, welche ihn die fernere Ausübung seines Berufs unmöglich machte und am 22. Februar 1882 seine Aufnahme in die Privat-Irrenanstalt des königlichen Sanitätsraths Dr. Hertz zu Bonn veranlaßte. Dort hat er sich nach Ueberwindung einer lebensgefährlichen Erkrankung des Gehirns zwar körperlich nach und nach erholt, ohne daß es jedoch bis jetzt zu einer Besserung seines krankhaften Geisteszustandes gekommen ist. Da Professor Winnecke bereits in den 1860er Jahren von einem ähnlichen Gemüthsleiden – melancholischer Verstimmung mit ängstlichen Wahnvorstellungen – befallen, aber nach einem Zeitraum von fünf Jahren aus der Hertz'schen Anstalt, in welcher er auch damals schon behandelt wurde, geheilt entlassen worden war, so erschien auch in dem vorliegenden Erkrankungsfalle die Hoffnung auf einen günstigen Ausgang nicht von vorn herein ausgeschlossen. Es wurde deshalb dem Observator der Sternwarte die Vertretung des Professors Winnecke im Lehramte und in der Direktion der Sternwarte bis auf weiteres übertragen, wodurch freilich für die Begabung und Thätigkeit dieses bedeutenden Instituts, welches nach dessen Plänen und unter seiner Leitung erbaut und mit wissenschaftlichen Instrumenten ausgerüstet, dazu bestimmt war, unter den deutschen Sternwarten eine hervorragende Stellung einzunehmen, ein nur nothdürftiger Ersatz geschaffen werden konnte.

Als unter diesen mißlichen Verhältnissen bereits vier Jahre verflossen waren, und die von Zeit zu Zeit über den Verlauf der Krankheit des Professors Winnecke in vertraulicher Weise eingezogenen Nachrichten eine Wendung zum Besseren nicht erkennen ließen, uach durch die für Ostern laufenden Jahres erfolgte Berufung des stellvertretenden Observators an die Universität Göttingen eine weitere Verschlimmerung der Sachlage in Aussicht stand, trat die unabweisliche Nothwendigkeit heran, von längerem Zuwarten abzusehen und die schwebenden Fragen einem Abschlusse entgegenzuführen. Sanitätsrath Dr. Hertz wurde zur Abgabe eines eingehenden Gutachtens aufgefordert, welches derselbe am 13. Januar dieses Jahres erstattete. Am Schlusse desselben heißt es: "Aus dem Vergleiche beider" (das heißt der günstigen und der ungünstigen) "Symptome geht hervor, daß die Aussichten auf die Wiederherstellung auch im Verlaufe der Zeit eher gesunken als gestiegen sind. Wenn gerade bei diesem rationell veranlagten und mit hochpotenzierten spekulativen Geisteskräften ausgestatteten Manne auch manche außergewöhnliche Leistung in Bezug auf eine restitutio in integrum *gestörter Gehirnfunktionen erwartet werden kann, woran sich die Hoffnungen knüpften, so ist es doch bei diesen problematischen Voraussetzungen unmöglich, daran zu denken, daß er besten Falles ein öffentliches, für ihn mit angewiesenen Pflichten verbundenes Amt antreten könnte. Schon die ärztliche Vorsicht würde ihm dieses für immer verbieten." Mit diesem Urtheil hat sich der Direktor der psychiatrischen Klinik der hiesigen Universität, Professor Dr. Jolly, welchem das Hertz'sche Gutachten zur Prüfung vorgelegt worden war, unter eingehender Begründung seiner Ansicht vollkommen einverstanden erklärt.*

Auf eine Wiedergenesung des Professors Dr. Winnecke ist hiernach keine Hoffnung. Die Interessen der Universität erheischen aber dringend eine Wiederbesetzung des verwaisten Lehrstuhles der Astronomie. Die Emeritierung des Professors Dr. Winnecke müßte hiernach in Aussicht genommen werden. Er selbst hat sich mit derselben einverstanden erklärt und zwar nach dem Zeugnisse des Arztes im vollen Bewußtsein von der Wichtigkeit und Tragweite dieses Schrittes[6].

[The full professor in the Faculty of Mathematics and Sciences of the Emperor Wilhelm University Strasbourg and director of the University Observatory, Dr. Winnecke, suffers since the beginning of 1882 from a melancholia, which made the further practice of his profession impossible, and instigated, on 22 February 1882, his admittance in the private lunatic asylum of the royal Sanitätsrath Dr. Hertz in Bonn. There he has gradually recovered physically after overcoming a perilous sickness of the brain, however, until now no recovery of his pathological mental state has occurred. Since Professor Winnecke has been stricken in the 1860s by a similar melancholia with pathological crazy notions, but had been released after five years from Hertz' asylum, where he was also treated at that time, hope for a favorable outcome did not immediately appear to be excluded in the present case as well. The observer [Observator] was asked to act as a substitute of Professor Winnecke for the teaching and the direction of the observatory. Thereby, however, only a scanty replacement was found for the potential and the activity of this important institute, which was built according to his [Winnecke's] plans and under his direction and was equipped with scientific instruments, so that it was destined to fill an excellent position among German observatories.

When under these miserable circumstances already four years had passed, and the confidentially acquired news about the course of Prof. Winnecke's illness did not indicate a turn for the better, and also the call of the substituting observer to Goettingen University indicated an additional turn for the worse, it was felt absolutely necessary not to wait any longer, and to put an end to the pending questions. Sanitätsrath Dr. Hertz was asked to give a detailed judgement, which he reported on January 13. Its concluding lines read: "comparing both (favorable and infavorable) symptoms, it follows that the expectations for a recovery have declined in the course of time. When just in this person, who is rationally based and endowed with high-potential speculative mental forces, quite some extraordinary power with respect to a *restitutio in integrum* of disturbed cerebral functions can be expected, it is impossible to think, with these problematic suppositions, that even in the most optimistic case he can fill a public position which is connected with his duties. Already medical precaution would forbid it forever." The director

[6]Geheimes Staatsarchiv PK, I Rep 89, No. 21694, p. 47.

Figure 3. Two acting directors of Strasbourg Observatory, Wilhelm Schur and Hermann Kobold (Shane Archives and Kobold estate).

of the psychiatric clinic of this university, Professor Dr. Jolly, has examined Hertz' report, and has agreed completely to the judgement.

Therefore there is no hope for a recovery of Professor Dr. Winnecke. The interests of the university require an urgent filling of the vacant professorial chair of astronomy. The retirement of Professor Dr. Winnecke has thus to be contemplated. He has agreed to such a thing, according to the witness of the doctor, in full knowledge of the importance of this move.]

Winnecke's first two assistants were Wilhelm Schur (first assistant from 1873 onward) and Ernst Hartwig (second assistant from 1874 onward). On 1 July 1877, Schur was promoted to observer, and Hartwig to first assistant. From 1882, because of Winnecke's illness, Schur became acting director. Hartwig left in April 1884 to become astronomer in Dorpat. The replacement in the posts of assistants were Walter Wislicenus (1884-1889) and August Kaufmann (1884-1889). When Wislicenus was absent for half a year in 1885, Carl Stechert occupied the post of assistant (Stechert had obtained his PhD in Kiel the previous year, and later made his career from assistant worker to professor at the Maritime Observatory [Seewarte] in Hamburg).

As already mentioned in the preceding letter, Schur (Fig. 3) quit in April 1886 to become director of Goettingen Observatory. His post was filled by Hermann Kobold (Fig. 3), who had worked before at Nicolaus von

Konkoly's private observatory in O'Gyalla and with the German Commission for the observation of the Venus transits in Berlin, under the supervision of Arthur Auwers. Wislicenus had hoped to obtain the position of observer, but if there were any attempts from the side of the university in this respect, they did not bear fruit: Kobold was obviously the choice of the Berlin ministry, which was likely acting on the advice of Auwers. Now, freshly-appointed Kobold became acting director of Strasbourg Observatory. The search for a new director had not yet begun; it would turn out to be quite tiresome, and would take more than a year.

2. Consolidation: The Era of Ernst Becker

Several astronomers like Eduard Schoenfeld, director of Bonn Observatory, and Friedrich Tietjen, director of the Astronomical Computing Institute [Astronomisches Rechen-Institut] in Berlin, declined the offer to accept the Strasbourg Observatory directorship, and there were already plans to give the post to Karl Schering, a mathematical physicist at Strasbourg University, if another attempt would fail. This time, however, things turned out well. Ernst Becker accepted, and started work at the end of 1887. Becker, who was born in Emmerich, Rhineland, had been observer in Berlin between 1874 and 1883, where he carried out observations of the Astronomische Gesellschaft Catalogue project, zone $+20° \ldots +25°$, and subsequently director of the small Gotha Observatory. It perhaps remains to be added that Becker took part in the German expedition to observe the Venus transit of 1874 in Isfahan, Persia, as the only astronomer. It was mostly a photographic expedition, and its leader Gustav Fritsch, photographer and *Doctor medicinae*, constated that Becker – unknowingly – suffered attacks of epilepsy under stressful conditions (letter from Fritsch to Foerster, 29 December 1874; Foerster estate, archive BBAW). Becker (Fig. 4) would occupy the Strasbourg post for 22 years – the longest time of service of a German director.

Here is the official document of Becker's employment, addressed to the Emperor and King, and dated Strasbourg, 17 August 1887:

Auf Grund der Kabinetts-Ordre Eurer Majestät vom 31. Mai 1886 ist der ordentliche Professor in der mathematischen und naturwissenschaftlichen Fakultät der Kaiser-Wilhelms-Universität Straßburg Dr. Winnecke vom 1. Oktober 1886 ab emeritiert worden. Behufs Wiederbesetzung der von ihm bekleideten Professur für Astronomie nebst der Direktion der hiesigen Sternwarte sind seitdem mit einer Reihe von Gelehrten Verhandlungen gepflogen worden, ohne zum Ziele zu führen. Erst jetzt nach Jahresfrist ist es gelungen, in der Person des von der mathematischen und naturwissenschaftlichen Fakultät vorgeschlagenen Direktors der Sternwarte zu Gotha Dr. Becker einen geeigneten Ersatz für den Professor Dr. Winnecke zu gewinnen.

Dr. Ernst Becker hat seine Studien zu Berlin in Berlin gemacht und ist darauf

Figure 4. Ernst Becker (Courtesy Leiden Obs.).

an der Sternwarte zu Neufchâtel als Assistent und an den Sternwarten zu Leiden und Berlin als Observator thätig gewesen. Seit 1881 ist er Direktor der Sternwarte zu Gotha. Seine Arbeiten werden von den ersten Autoritäten des astronomischen Faches als tüchtige Leistungen bezeichnet. Es ist daher wohl zu hoffen, daß er auch den größeren Anforderungen, welche die Straßburger Sternwarte an ihren Direktor stellt, genügen wird. Als Dozent hat sich Dr. Becker allerdings noch nicht versucht. Allein da seine Fachgenossen auf Grund ihrer persönlichen Bekanntschaft mit ihm bestimmt annehmen, daß er die Gelegenheit, jüngere Leute in seine Wissenschaft einzuführen und für dieselbe heranzubilden, freudig und mit Eifer ergreifen und schnell die erforderliche Uebung im Doziren erlangen werde, so kann über jenes Bedenken hinweg gesehen werden.

Direktor Dr. Becker hat sich bereit erklärt, die erledigte Professur anzunehmen und bald nach dem Beginn des nächsten Wintersemesters anzutreten.

Eurer Majestät gestatte ich mir hiernach die allerunterthänigste Bitte vorzutragen: Allerhöchstdieselben wollen Allergnädigst geruhen, durch huldreiche Vollziehung der angeschlossenen Bestallung den Direktor der Sternwarte zu Gotha Dr. Ernst Becker zum ordentlichen Professor in der mathematischen und naturwissenschaftlichen Fakultät der Kaiser-Wilhelms-Universität zu ernennen[7].

[...on the basis of the cabinet ordre of your majesty of 31 May 1886, the full professor in the mathematical and scientific faculty of the Emperor Wilhelm University Strasbourg, Dr. Winnecke, has retired by 1 October 1886. In order to fill again his professorship for astronomy, as well as the directorship the observatory, negotiations have been carried out with a number of scholars, however, without success. Only after a year, it was possible to win a suitable replacement for Professor Dr. Winnecke, in the person of the director of Gotha Observatory, Dr. Becker, who has been proposed by the mathematical and scientific faculty.

Dr. Ernst Becker has carried out his studies in Berlin and thereafter was assistant at Neufchâtel Observatory, and observer at Leiden and Berlin. Since 1881, he is director of Gotha Observatory. His studies are judged as by the first astronomical authorities as sound achievements. One can therefore hope that he will suffice the higher demands which are posed by Strasbourg Observatory to its director. Dr. Becker has, however, no teaching experience. It is only because his colleagues assume with certainty, based on their personal acquaintance, that he will accept with zest the opportunity to introduce younger people into his science, and to educate them in it, and will quickly gain the necessary experience of teaching, one can manage to see over this scruple.

Director Dr. Becker has agreed to accept the vacant professorship and to be available soon after the beginning of the next Winter term.

Hereafter I dare to pose the submissive demand to your majesty: May it please your majesty to appoint the director of Gotha Observatory, Dr. Ernst Becker, to the post of full professor in the mathematical and scientific faculty of the Emperor Wilhelm University.]

The appointment of Becker was approved by the Emperor on 23 August 1887. Soon after Becker's arrival in Strasbourg on 1 December 1887, Wislicenus handed in to the faculty his habilitation thesis, in order to become private lecturer [Privatdozent]. This was a reason for Kobold to also hand in a habilitation thesis on the proper motions of fundamental stars, which was accepted in February 1888.

In 1889, the staff underwent major changes. The second assistant, Kaufmann, having passed the examination for higher instruction, left on 1 April to take up a position at the gymnasium in Altkirch (Alsace). Becker had found indications that Wislicenus was plotting against him, and dismissed him on 1 October. The positions of first and second assistant were filled by Max Arthur Zwink (Fig. 5) and Jacob Karl Ernst Halm. Zwink, born

[7]Geheimes Staatsarchiv PK, I Rep 89, No. 21694, p. 93.

Figure 5. Some assistants of Strasbourg Observatory: M. Zwink, B. Wanach, M. Ebell, W. Ebert, L. Carnera, and E. Jost (from PGAG 1904).

in 1858, had been assistant in Berlin from 1883-1888, and got his PhD in Kiel in 1888 on *Die Pendel-Uhren im luftdicht verschlossenen Raume mit besonderer Anwendung auf die bezüglichen Einrichtungen der Berliner Sternwarte* (obviously he had been in charge of Berlin Observatory's clocks in previous years). Halm, having worked at Kiel in the editorial office of the *Astronomische Nachrichten*, and graduated with a mathematical topic, had been recommended by its editor, Adalbert Krüger.

In 1888, Becker had started to carry out another Astronomische Gesellschaft project, the cataloguing of stars between $-1°\ldots-6°$ declination with the Repsold meridian circle, in which also all assistants were involved (Halm alone would carry out about a third of the observations). This project was published 18 years after its incipience (Becker 1906).

1892 was another year of major changes: Zwink apparently had also plotted against the director, and was dismissed by Becker. In Kobold's notes, we read that *between Becker and the servant of the Observatory, also a state of war had been declared...* The successor of Zwink was Berthold Wanach (Fig. 5). In Autumn, The Earl of Crawford had given his observatory to the state, and its director Copeland was looking for astronomers. When he asked Kobold to accept a position, the latter refused, but suggested Halm instead, who went as first assistant to the Royal Observatory at Edinburgh in 1895. The first assistant Wanach left in early 1898 to accept a position at the geodetic institute in Potsdam. His position was filled by Martin Ebell (1897-1902) (Fig. 5), who afterwards went to Kiel, and became a long-term employee of the editorial office of the *Astronomische Nachrichten.* His successor was Wilhelm Ebert (Fig. 5), who had got his PhD in 1895 in Munich under Seeliger, and had been an assistant in Paris and Kiel before. He already left at the end of 1902, first to Paris, then became observer in Berlin, and was killed in World War I. Luigi Carnera (Fig. 5) took over, but left already in September 1903, and the post remained unfilled for a year. Ernst Jost (Fig. 5), who had been assistant in Heidelberg and Gotha before, took over on 1 November 1904, and only left by the end of 1909, to become observer in Koenigsberg, where he died at the end of World War II.

Halm's successor in the post of second assistant was Karl August Ferdinand Necker (1895-1897), who died in a train accident in Egypt, where he had sought cure from tuberculosis. He was followed by Otto Tetens (1897-1902) (Fig. 6), who afterwards had a career as a geophysicist on Samoa, employed by the Goettingen Society of Sciences, and subsequently as an astronomer in Kiel and as a scientist at the meteorological observatory Lindenberg near Berlin, and by Harry Meyer (1902), who later became a teacher in navigation. In the next two years, Berthold Cohn, Karl Schiller (Fig. 6) and Erich Redlich served as assistants; Cohn would later become the observatory computer for several decades, Schiller – after some stays at

Figure 6. Some assistants of Strasbourg Observatory: O. Tetens (from PGAG 1904), K. Schiller, W. Hassenstein, and V. Heinrich (from PGAG 1931).

other observatories – would become assistant up to the end of the German era, Redlich would become assistant at Kiel Observatory from 1909-1914, a volunteer at the outbreak of war, and would be killed on the Western Front in March 1918.

The position of second assistent was afterwards filled by Felix Biske (1905-1907) and Julius Liebmann (1907-1913). Biske, originating from Płońsk, had obtained his PhD in Berlin in 1901 with an *attempt to apply hydrodynamic investigations to solar prominences.* In 1902 and 1909, he worked at Zuerich Observatory, his last astronomical contribution, published in the *Astrophysical Journal*, on diffraction phenomena during eclipses is from 1913, was written when Biske was working in the Physics Department of the Imperial University of Warsaw; his later fate is unknown.

Julius Liebmann obtained his PhD at Berlin University in 1905 under W. Foerster on an instrumental-astrometric topic. After leaving Strasbourg, he became an observer at Berlin Observatory, but was killed in Belgium in the first days of World War I.

Kobold remained as an observer in Strasbourg (Fig. 7) until April 1902. On 11 March 1900, he was also appointed extraordinary [außerordentlicher] professor. During his whole stay in Strasbourg, his salary remained fixed at 5000 Marks (annually), because salary increases because of longer service time did not exist in Alsace-Lorraine. Therefore, after many fruitless applications for director's posts in Gotha, Hamburg, and Goettingen, Kobold accepted the position of observer and extraordinary professor at Kiel University, where his salary was almost twice as high. From time to time, he reported in the annual reports on progress in the reduction of his Strasbourg solar heliometer observations, but even at the time of his death, in 1942, definitive results had not yet been obtained. Hellerich, at that time director of Strasbourg Observatory, agreed to complete the project, but the material possibly disappeared in the early post-WWII period (Duerbeck 2005b).

After Kobold's move to Kiel, Carl Wirtz was appointed observer (temporary in 1902, definitive in 1903). Also his starting salary remained modest: 5600 Marks *per annum*, and a "price increase supplement" which amounted to 1170 Marks at the end of the war.

Let us now focus our attention again on Wislicenus, who had been a private lecturer since early 1889, and not attached anymore to the observatory since late 1889. Nevertheless, he was permitted to use the small 162mm refractor for his studies of surface structures on Mars and for brightness measurements on lunar surface features. In 1890/91, he had already started to give lectures on "astrophysical" topics (astrophysics of the solar system, spectroscopy of the sky), which would culminate in 1899 in his small book *Astrophysik*. In 1894, he was appointed extraordinary professor. Soon after, in mid-1895, the faculty applied through the curator to the Imperial Governor [Statthalter] to establish an extraordinary professorship of astrophysics. This might be interpreted as an attempt to install the "new astronomy" in Strasbourg, as it is done by Wolfschmidt (2005), but the main reason was undoubtedly the move of the faculty to provide a paid position for Wislicenus. But it was declined later in the year by the *Statthalter*. Wislicenus started thinking on another topic he could excel in, and this finally culminated in his editing of an annual bibliography of astronomy, starting from 1899 (see also Duerbeck 2005a).

Becker's reign is marked by the beginning of the series of publications of

Figure 7. Strasbourg Observatory in the 1890s, in the foreground a son of Hermann Kobold (Kobold estate).

the observatory. Three volumes appeared under his directorship; the first two were prepared with the assistance of Kobold, the third one (in two parts) by Wirtz. He was eager to see the plans of his predecessor continued and its results published; he filled the new positions with talented, although not necessarily "first choice" people who were willing to carry out their daily (and nightly) duties, and could carry out some own research. Additional observing programs were the monitoring, by means of the altazimut telescope, of the polar altitude, to follow its newly-discovered variations, the monitoring of the equatorial and polar diameter of the sun, carried out with the heliometer, and observations of comets down to the limit of visibility, by means of the large refractor.

Besides an extensive handbook article (Becker 1899) on *Micrometers and micrometer measurements*, Becker (1906) also arranged and prepared the Astronomische Gesellschaft Catalog (AGK1) for the zone $-2°\ldots -6°$, containing the positions of 8204 stars, determined with the Repsold meridian circle, based on observations by Halm, Zwink, Wislicenus, Becker, Wanach and others in the years 1888-1893, 1895, 1903-1905; besides this, he also published his own Berlin meridian circle observations of 1879-1883, which covered the zone $+20°\ldots + 25°$ (Becker 1895). The determination of the longitude difference between the Grand-Ducal Observatory (Astronomical Institute) near Heidelberg and the Imperial University Observatory in Strasbourg, carried out in 1903, was prepared and edited by Becker & Valentiner (1906).

Another one of his projects, presumably requested by the government in 1898, was the carrying out of relative gravity observations at various points in Alsace-Lorraine that should complement those made in the Grand-Duchy of Baden, east of the Rhine. In six campaigns during the years 1900-1905, measurements were carried out in 44 locations. Besides two geodesy scientists from Karlsruhe and Darmstadt, his observatory collaborators Kobold, Carnera, Jost and Schiller participated in the observations. The final report was written after his retirement, and was delayed by sickness; actually, the preface was written three months before his death (Becker 1912).

During the time of Schur, Kobold and Becker, six PhD theses were completed: by Walter Wislicenus (1886), Friedrich Ristenpart (1892), Berthold Cohn (1897), Henry Allen Peck (1897), Leopold Courvoisier (1900) and Hans Rosenberg (1905). Among the PhD absolvents, most made their way in astronomy: Walter Wislicenus, who obviously had chosen the topic on his own, has already been mentioned. Friedrich Ristenpart (Fig. 8), whose topic may have been inspired by Kobold, afterwards moved to Heidelberg and then became a scientific employee at the Berlin Academy (probably also because of the association between Kobold and Auwers), to start the project *History of the starry sky* [Geschichte des Fixsternhimmels], a collection of all stellar positions, derived from transit observations or equivalent methods, from 1750 to about 1900. This project, first supervised by Arthur Auwers, was some sort of precursor of the *Centre des Données Stellaires.* In 1908, ten years after its incipience, Ristenpart went to Santiago de Chile as director of the Chilean National Observatory, where he committed suicide five years later when his contract was not prolonged. The project, however, survived Friedrich Ristenpart, Arthur Auwers, the German Empire, the Weimar Republic, and the Third Reich: the last of the 24 volumes of the *Geschichte des Fixsternhimmels* appeared in 1964.

Henry Allen Peck (Fig. 8) spent the rest of his life at Syracuse, NY, USA, first as a mathematics instructor and later professor and director of the Syracuse University Observatory. Leo Courvoisier (Fig. 8) became a specialist in astrometry at Berlin-Babelsberg Observatory. Hans Rosenberg, after some astrophysical work at his private observatory in Tübingen, was appointed director of the Kiel Observatory. After his dismissal by the Nazis, he became director of Istanbul Observatory. Berthold Cohn (Fig. 8), who was born in 1870 in Rawitsch (poln. Rawicz, Wielkopolska) had studied in Wroclaw [Breslau], Basel, Paris and Strasbourg. He also acquired a high degree of Jewish learning, and published various articles in the *Jahrbuch der Jüdisch-Literarischen Gesellschaft* (Yearbook of the Jewish Literary

Figure 8. Some PhD alumnae of Strasbourg Observatory: Ristenpart, Cohn, Peck, Courvoisier (from PGAG 1904), Semenov (from PGAG30), and Messow (Schramm 1996).

Society, Frankfurt). He was second assistant in 1903-1904, ceased to publish in astronomical journals in 1906, but remained as a scientific assistant worker [calculator] at the Observatory. In 1918, his *Almanach perpetuum des Abraham Zakuto*, an edition and German translation of a manuscript of Abraham Sacut or Zacuto (ca. 1450 – ca.1510), appeared (Cohn 1918) – a fact listed also in the Observatory's annual report that Cohn had made these studies *privatim*. Since he did not qualify as a *Boche* [Kraut], he kept his job also under French rule: *M. Cohn, ancient assistant calculator of the observatory, having become Polish by virtue of the peace treaty, remained employed at the observatory*, as stated in the text of the new organisation of the observatory (Esclangon 1926). Cohn died in Strasbourg in May 1930; at the time of his death, he was still a calculator at the observatory, as well as a member of the *Astronomische Gesellschaft*.

After the long term of 22 years, where he had not only published many catalogues, but also supervised PhD theses, and had instructed many students (perhaps the most notable was Karl Schwarzschild between 1891-1893), Becker sought retirement. The official statement, addressed to the Emperor and King, dated Strasbourg, 11 November 1908, reads:

Der ordentliche Professor in der mathematischen und naturwissenschaftlichen Fakultät der Kaiser-Wilhelms-Universität Straßburg Dr. Ernst Becker hat seine Emeritierung zum 1. April 1909 erbeten.

Gemäß §44 des Statuts der Universität kann die Emeritierung eines Professors mit seiner Einwilligung jeder Zeit erfolgen; berechtigt, sie zu verlangen, ist der Professor nur, wenn er das 65. Lebensjahr vollendet hat oder wenn ihm die Leitung einer wissenschaftlichen Anstalt entzogen wird.

Professor Dr. Becker ist am 11. August 1843 geboren, steht somit im 66. Lebensjahre.

Euere Majestät bitte ich alleruntertänigst, durch huldreiche Vollziehung des im Entwurfe angeschlossenen Erlasses die Emeritierung des Professors Becker zum 1. April 1909 Allergnädigst genehmigen zu wollen[8].

[The full professor in the mathematical and scientific faculty of the Emperor Wilhelm University Strasbourg, Dr. Ernst Becker, has asked for his retirement on 1 April 1909.

According to §44 of the statute of the University, the retirement of a professor can be carried out any time; he is only entitled to ask for it when he has completed the 65th year or when he is deprived from the directorship of a scientific institution. Professor Dr. Becker was born on 11 August 1843, and is thus in his 66th year.

I ask your majesty submissively, by benevolent execution of the decree attached in draft form, to grant the retirement of Professor Becker on 1 April 1909.]

The answer, dated 18 November 1908, and addressed "To my Governor in Alsace-Lorraine", reads:

Auf Ihren Bericht vom 11.ten November dieses Jahres will Ich Mich damit einverstanden erklären, daß der ordentliche Professor in der mathematischen und naturwissenschaftlichen Fakultät der Kaiser-Wilhelms-Universität Straßburg Dr.

[8]Geheimes Staatsarchiv PK, I Rep 89, No. 21696, p. 96.

Ernst Becker mit dem im Statut dieser Universität vorgesehenen Wirkungen zum 1. April 1909 ab emeritiert wird [9].

[As of your report of November 11 of this year I will declare myself in accordance with the fact that the full professor in the mathematical and scientific faculty of the Emperor Wilhelm University Strasbourg, Dr. Ernst Becker, retires by 1 April 1909, with the consequences foreseen in the statute of this university.]

3. Interlude: The Beginnings of Seismology

In the 1892 annual report of Strasbourg Observatory, Becker wrote: *Finally I mention that last year, upon request of Dr. von Rebeur-Paschwitz, I have installed a horizontal pendulum, which was previously functioning for some time at the Wilhelmshaven Observatory, in the basement of the meridian building. It is put up on a massive stone console embedded in the large pillar of the Cauchois passage instrument... The apparatus has functioned well for nearly 3/4 of a year without interruption, thanks to the careful surveillance of the doorman and calculator C. Sabel, who also carries out the development of the photograms with good talent.*

Ernst von Rebeur-Paschwitz (1861-1895; Fig. 9), an astronomer by education, had developed a very sensitive self-registering horizontal pendulum, and carried out measurements in Wilhelmshaven, Potsdam, and Puerto Orotava (Tenerife). Rebeur-Paschwitz's original intent was to register the tides of the solid earth (the so-called "moon wave"), and he looked for low-frequency signals. But on 17 April 1889, he recorded in Potsdam and Wilhelmshaven oscillations which occurred almost simultaneously with an earthquake in the vicinity of Tokyo, which had an unusually low frequency: *We may therefore safely conclude that the disturbances noticed in Germany were really due to the volcanic activity which caused the earthquake of Tokyo* (Rebeur-Paschwitz 1889). This was the first recording of a *teleseism.*

The Wilhelmshaven instrument was refurbished by the Repsold Company in Hamburg, and installed at Strasbourg Observatory. Since von Rebeur-Paschwitz was at that time already handicapped by illness, he was grateful to Becker to arrange for the functioning of the instrument. The Potsdam/Puerto Orotava instrument was sent to the Russian Naval Observatory at Nikolayev, where its director Ivan Kortazzi carried out parallel observations (and later purchased the instrument for 718 Marks). Results for 1892, based on the Strasbourg and Nikolayev instruments, were published (von Rebeur-Paschwitz 1892). Especially the *earthquake in Beludschistan* of 20 December 1892, an event with a magnitude of 6.8 in the present Afghanistan-Pakistan border region, was recorded both in Strasbourg and Nikolayev (Rebeur-Paschwitz 1894).

[9] Geheimes Staatsarchiv PK, I Rep 89, No. 21696, p. 97.

Figure 9. Ernst von Rebeur-Paschwitz (courtesy of Universitätsarchiv Tübingen (UAT 209/108)).

In a detailed paper on observations with the horizontal pendulum at Strasbourg Observatory, von Rebeur-Paschwitz (1895a) showed that seismic waves propagate on paths along the earth's surface (surface waves) and through the earth's sphere (body waves) with different velocities, and thus such waves are best suited to investigate the earth's interior.

Rebeur-Paschwitz, terminally ill with tuberculosis, realized the necessity for international cooperation, and for the Sixth International Geographic Conference which took place in London in 1895, he suggested the establishment of an international system of earthquake stations: "all stronger earth- and seaquakes, wherever they will take place, must leave their traces on the photograms of the planned stations" (Rebeur-Paschwitz 1895b). The Strasbourg geography professor Georg C.K. Gerland (1833-1919), who had purchased Rebeur's Strasbourg instrument for his own monitoring program, presented Rebeur's ideas at the conference and achieved a widely supported resolution. In spite of this, there was no imminent international collaboration, and the Englishman John Milne established a network of stations with simply-to-use horizontal pendulums in the British colonies.

During the Seventh International Geographic Conference, which took place in Berlin in 1899, Georg Gerland again presented the ideas, and an *International Permanent Commission for Earthquake Research* was founded. In the same year, an imperial main station for seismic research [Hauptstation für Erdbebenforschung] was established in Strasbourg. Consequently, in 1901, the *First International Conference on Seismology* took place in Strasbourg, which was followed by a second one in 1903. A group photograph of the first one shows the members of the observatory Ebell, Kobold and Tetens as participants on the conference group photograph (Schweitzer 2003). The conference proceedings were published in supplementary volumes of Gerlands Beiträgen der Geophysik (1902 and 1904) (Kertz 1999).

At the 1901 conference, it was decided to install a *Central Bureau*, whose main office should be in Strasbourg. It should collect and disseminate the reports of all seismic stations, and should coordinate seismic research. However, both national and international obstacles had to be removed (for a detailed discussion see Cremer 2001). The activities were basically boycotted by French scientists, British participants were also somewhat reserved since they feared they would impinge upon national undertakings, and even the German geophysics pioneer Emil Wiechert went his own way. Nevertheless, Strasbourg indeed became a center of seismic research; a first circular of the Central Bureau by Gerland (1906) was published in the *Astronomische Nachrichten*. Some years later, at the time of Gerland's retirement in 1909, there were problems to fill the positions of the professor of geography and of the director of the seismic station, which should not be run anymore by the same person, and the question arose whether the "Reichsland" or the Prussian state should finance the latter one. Again, a few years later, World War I disturbed international relations, although the Central Bureau tried to fulfill its duties according to the rules. Nevertheless, its director Oskar Ernst August Hecker (1864-1938) had to leave Strasbourg on 6 January 1919, and the geophysics professor Edmond Rothé (1873-1942) was put in charge of the university institute, while the central office was run for a few months by Lieutenant Baldit, but then also taken over by Rothé.

4. Twilight: The Era of Julius Bauschinger

At the retirement of Ernst Becker, it was not so difficult to find a successor. Julius Bauschinger (Fig. 10), a professor at Berlin Observatory, and head of the Astronomical Computing Institute [Astronomisches Rechen-Institut], became director in April 1909. The official letter, addressed to the Emperor and King and dated Strasbourg, 25 January 1909, reads:

In der mathematischen und naturwissenschaftlichen Fakultaät der Kaiser-Wilhelms-Universität Straßburg wird durch die Emeritierung des Professors Dr.

Figure 10. Julius Bauschinger – as a student, as head of the Berlin Astronomical Computing Institute, and as the retired director of Leipzig Observatory (Shane Archives, PGAG 1904 & PGAG 1931).

Becker, welche Euere Majestät durch Allerhöchsten Erlaß vom 18. November 1908 zu genehmigen geruht haben, vom 1. April ab die ordentliche Professur für Astronomie zur Erledigung kommen. Zur Wiederbesetzung derselben hat die genannte Fakultät den ordentlichen Professor an der Universität zu Berlin Dr. Bauschinger in Vorschlag gebracht.

Julius Bauschinger, geboren in Fürth, Kreis Mittelfranken, Königreich Bayern, am 28. Januar 1860, katholischer Konfession, erhielt seine Universitätsbildung in Berlin und München, promovierte im Dezember 1883 in München, wurde 1886 Observator an der Sternwarte daselbst und habilitierte sich 1888 an der dortigen philosophischen Fakultät. Ostern 1896 folgte er einem Rufe als ordentlicher Professor und Direktor des astronomischen Recheninstituts an die Universität in Berlin. Er ist von der praktischen Astronomie ausgegangen und hat auf diesem Gebiete vediente Arbeiten geliefert. Sein Amt als Direktor des Recheninstituts verlangt eine intensivere Beschäftigung mit der rechnenden Astronomie und hier hat Bauschinger Hervorragendes geleistet. Die Herausgabe des astronomischen Jahrbuchs besorgt er mustergültig; er leitet die Arbeiten des Bureaus ausgezeichnet und entwickelt darin selbst eine große Thätigkeit. Auch literarisch ist er vielfach thätig und sein jüngst erschienenes Werk "Die Bahnbestimmung der Himmelskörper" ist eines der besten Lehrbücher der theoretischen Astronomie.

Professor Bauschinger ist gemäß seiner Erklärung bereit, die zur Erledigung kommende Professur, mit der die Direktion der Universitätssternwarte verbunden ist, zum 1. April 1909 zu übernehmen. Seitens des königlich Preussischen Ministeriums der geistlichen, Universitäts- und Medizinalangelegenheiten steht nach eingezogenen Erkundigungen der Berufung des Genannten zu dem angegebenen Zeitpunkte kein Bedenken entgegen.

Euere Majestät bitte ich hiermit alleruntertänigst, Allerhöchstdieselben wollen Allergnädigst geruhen, durch huldreiche Vollziehung der angeschlossenen Bestallung den ordentlichen Professor Dr. Julius Bauschinger in Berlin zum ordentlichen Professor in der mathematischen und naturwissenschaftlichen Fakultät der Kaiser-

Wilhelms-Universität Straßburg zu ernennen[10].

[Because of the retirement of Professor Dr. Becker, granted by your majesty's most supreme decree of 18 November 1908, the full professorship for astronomy in the mathematical and scientific faculty of the Emperor Wilhelm University Strasbourg becomes vacant on 1 April. The said faculty has proposed to fill this position with the full professor at Berlin University, Dr. Bauschinger.

Julius Bauschinger, born in Fuerth, in the region of Middle Franconia, Kingdom of Bavaria, on 28 Januar 1860, of catholic confession, obtained his university education in Berlin and Munich. He obtained his PhD in Munich in December 1883, became observer at Munich observatory in 1886, and got his habilitation in 1888 at the philosophical faculty in Munich. On Easter 1896, he was appointed full professor at Berlin University and director of the Astronomical Computing Institute. He started his career in observational astronomy and has produced distinguised studies. His post as director of the Computing Institute demands a more intense occupation with computational astronomy, and here Bauschinger has produced excellent work. His edition of the *Astronomical Yearbook* is exemplary; he guides the works of the office excellently and is himself very much involved in it. He has written profusely, and his recently issued book *Orbit determination of celestial bodies* is one of the best textbooks of theoretical astronomy.

Professor Bauschinger has declared to accept the vacant professorship, which is combined with the directorship of the university observatory, by 1 April 1909. On behalf of the Royal Prussian Ministry of Religious, University and Medicine Affairs, there is no objection to recruit the said person at the given time, according to collected information.

Herewith I ask your majesty most submissively, your most supreme excellency may please most graciously, by benevolent execution of the enclosed appointment, to appoint the full Professor Dr. Julius Bauschinger in Berlin to full Professor at the faculty of mathematics and sciences of the Emperor Wilhelm University Strasbourg.]

This was agreed upon on 1 February 1909. There is hardly anything to add. As a young student, Bauschinger was able to participate as an assistant in the 1882 Venus transit expedition to Hartford, Connecticut. His dissertation dealt with investigations on the motion of the planet Mercury. His activities in Berlin lead to the independence of the Astronomical Computing Institute, and to an addition of two budgetary posts. Nevertheless, after 13 years, he wanted to start a new life in a still quite powerful observatory, in a middle-size city, in a house in the middle of the observatory gardens.

New programs were developed for the refractor and the meridian circle: a photometry of 900 nebulae, and a survey of 1000 double stars for the bright nights. This program was carried out by Wirtz. The Repsold meridian circle, equipped with an impersonal micrometer, was used for a program of absolute positions of 1480 stars of 8-9th mag with very small proper motion, including solar observations and day and night measure-

[10]Geheimes Staatsarchiv PK, I Rep 89, No. 21696, p. 103.

ments of brighter fundamental stars. The observations were carried out by Walter Hassenstein, Julius Liebmann, and later by Karl Schiller.

In Summer 1912, the government requested to the observatory to determine the latitude of 22 geodetic points in Alsace-Lorraine. The observations were carried out by Bauschinger and Wirtz.

Under Bauschinger's directorship, Volumes 4 and 5 (both issued in two parts) of the Observatory Annals appeared, although the last volume was only issued in 1923/26, when Strasbourg Observatory was run by the French, and Bauschinger was director in Leipzig.

In Bauschinger's time, seven PhD theses were completed, by Heinrich Plate (1910), Erich Redlich (1911), Leonid Semenow (1911), Benno Messow (1913), Anton Tschetschott (1913), Martin Matzdorff (1914) and Carl Schumacher (1917). Semenow (Fig. 8) and Tschetschott returned to Russia; the first one became director of the Nikolayev Naval Observatory in 1926. Messow (Fig. 8), a Berlin student, making use of Potsdam photographic material, could obviously complete his thesis in Strasbourg after the death of his Berlin advisor, the astrophysicist Julius Scheiner (Bauschinger had been one of Messow's teachers in Berlin); since 1900, Messow had been assistant in Hamburg, and later became an observer at Hamburg Observatory. Schumacher, who was a teacher by profession, completed a thesis on a historical topic, and afterwards returned to school service; he died in 1929 as a secondary school teacher in Sterkrade (Rhineland).

The posts of director, observer and assistant remained almost unchanged till the end of the German era. After the departure of Ernst Jost who was appointed observer in Koenigsberg, Walter Hassenstein (Fig. 6) became first assistant in the years 1910-1918. After the departure of the secondary assistant Julius Liebmann (1907-1913), Karl Schiller (Fig. 6) (1913-1919) took his place. Berthold Cohn (Fig. 8) kept the position of "scientific assistant worker" (calculator). Extraordinary assistants were Paul Pummerer (1913), Martin Matzdorff (1914), and for a short time Vladimir Heinrich (Fig. 6) from Prague (1914), who already had spent some time at Strasbourg Observatory in 1908.

During wartime, the director was drafted immediately, he became officer and later on captain [Hauptmann] in the surveillance of the garrison; he also served as director of the censor's office for telegrams, and kept the directorship of the observatory. Matzdorff was killed at Yper in November 1914; Bauschinger wrote in the obituary *busy with sending out his just printed thesis, he encountered the outbreak of the war; without a moment's reflection he volunteered with the army, and only too soon the love for his country was sealed with his death.* The second assistant Pummerer was drafted in 1914, the doorkeeper/calculator Sause in 1915, the observer Wirtz and the assistant Hassenstein in 1916. In 1917, Schiller replaced Wirtz as observer.

Berthold Cohn kept the post of calculator, two ladies served as assistant calculators.

Pummerer worked at an airfield for kites [Felddrachenstation]. Wirtz first belonged to the people's army [Landsturm] and later on in the trigonometric department of the war headquarters [Großer Generalstab] in Berlin, while Hassenstein fought in Poland. While lectures were given intermittently, the observing program of the observatory became very much restricted during the war. The university served as a hospital for soldiers, soldiers camped in the observatory garden, or found shelter in the refractor building. At the end, only the meridian circle was kept active, the assistant Schiller acting as observer.

On 22 November 1918, the observatory was taken over by the French, and almost all employees of the university had to leave France. Bauschinger, the director, could only take with him his own observations, but not the meridian circle observations for the catalog of circumpolar stars, started already under Becker, and the observations of the faint small proper motion stars, started under his directorship. After some months of uncertainty that were spent in his home town Munich, Bauschinger was appointed director of the Leipzig University Observatory on 1 April 1920, taking the post from Heinrich Bruns, who had died on 23 September 1919. Bauschinger, already 60 years of age, served another 10 years as the director of Leipzig Observatory.

The observer and professor Wirtz obtained an equivalent position in Kiel (see also Duerbeck & Seitter 2005), Hassenstein became assistant and later observer in Potsdam. Schiller could remain in Strasbourg a few more months to continue the regular work, including the meridian circle observations. He was actually the only astronomer at the observatory until May 1919, when André Danjon assumed his work, and definitively left in August (Esclangon 1926); he became an assistant at Leipzig Observatory in mid-1920. Pummerer apparently found a position in the Maritime Observatory [Seewarte] in Hamburg, research of high wind measurements on a scientific aviation trip to Rio was published by him in 1939.

5. The Strasbourg-Leipzig Connection

As already mentioned, Bauschinger became director of Leipzig Observatory in 1920, and a few months later, when an assistant withdrew because of illness, Schiller was appointed assistant on 1 July 1920, and was promoted to observer in 1928. He busied himself with photometry, but was mainly occupied with meridian circle observations and reductions. He also worked on his Strasbourg observations of the star BD +89°1 (Polarissima), a project that appears in the annual reports between 1924-1929 and was taken up and published in 1937 (Schiller 1937).

Bauschinger was able to edit the fifth volume of the German Strasbourg Annals in two parts; they appeared in 1923 and 1926 and contained, besides double star observations made by Wirtz, also meridian circle results of stars, the sun, moon and planets, based on observations made until 1893. Bauschinger retired in 1930, and died in 1934.

In 1936, it was suggested to Josef Hopmann, the new director of Leipzig Observatory, to take over the reductions of Strasbourg meridian circle observations of the years 1893 to 1919, because of Bauschinger's and Schiller's earlier connection with the material. The Leipzig annual report reads: *After corresponding negotiations, the present director of Strasbourg Observatory, Prof. Dr. Danjon, as well as the French government have agreed that these meridian files should be sent [to Leipzig] as a loan for a number of years.*

The material consisted of two major sets, a catalog of almost 2000 circumstellar stars brighter than 7.5 mag between declination +60° and the north pole, observed between 1893-1909, and a catalog of almost 1500 fainter stars (7.5-8.5 mag) of small proper motion between declination −10° and the north pole, observed between 1910 and 1919. Both sets were fundamental observations, since they included solar observations in both coordinates. Reductions of the second set started in 1938 under the supervision of Schiller, and the subsequent annual reports indicate good progress, although they were slowed down in 1941. In 1942, it turned out that during wartime 1915-1916 the registration strip readings of the right ascension were not read and written down to the files, and the paper strips themselves had been destroyed *by French authorities* during the evacuation of Strasbourg Observatory in 1939. Besides this drawback, which meant that no right ascensions could be derived for 1915-1916, it appears that the reductions progressed well. However, Leipzig Observatory was hit by bombs on 4 December 1943, instruments and the library, already packed for safe storage, were destroyed, and working conditions became poor; part of the personnel was carrying out calculation for the armed forces.

An overview of the philosophical faculty of 27 June 1945 indicated that Hans Naumann (1883-1953) and Karl Schiller (1882-1979) still carried out computations for the Strasbourg observations, but Schiller left Leipzig in June 1945 after the withdrawal of US troups, Naumann became seriously ill and died in 1953, the Observatory was understaffed and basically closed in the mid-1950s, while astronomy classes for teachers were given by guest lecturers from other universities (Illgauds & Münzel 1995).

A brief note by Kohlschütter (1948) does not indicate whether the project was still alive after the war; nothing was published, and the actual whereabouts of the observational material are unknown: was it destroyed during wartime, was it taken by Schiller, was it still kept at the observatory, or even returned to France, as promised? As with the case of Kobold's heliometer measurements, the traces vanish in the early post-war period (see also Duerbeck 2005b).

Schiller, who had moved to the vicinity of Frankfurt in 1945, took over the provisional directorship of Frankfurt Observatory in 1947, when he was already 65, and he retired in 1960 (Bethge & Klein 1989). Forgotten by his colleagues, he died in 1979, almost a centenarian, in his home near Frankfurt – the last survivor of the Imperial Observatory.

6. Sources and Acknowledgments

Documents used for this study are kept in the Geheimes Staatsarchiv Preußischer Kulturbesitz (Berlin) and the Archiv der Berlin-Brandenburgischen Akademie der Wissenschaften (Berlin), and I thank these organisations for granting access to the documents. The information on Rebeur-Paschwitz is taken to a large extent from the detailed studies of Schweitzer (2003, 2004); supplementary information was given by G. Pinigin (Nikolayev Observatory). Additional details on the activities of Gerland, the Strasbourg seismic station and the Central Bureau and their history in German and French times are found in Cremer (2001). I am grateful to the Shane Archives (University of California at Santa Cruz) for the photographs of Bauschinger, Becker, Elkin, and Schur, to the Universitätsarchiv Tübingen for the photograph of von Rebeur-Paschwitz, to F. Schmeidler (München) for granting access to his 1904 Porträtgallerie, to R. and H. Vollmer (Altenholz) for the photographs of Kobold, as well as for granting access to Kobold's autobiography. I also thank G. Münzel (Leipzig) for useful informations.

References

1. Auwers, A. 1887-1898, Die Venusdurchgänge 1874 und 1882, Bericht über die deutschen Beobachtungen, Im Auftrage der Commission für die Beobachtung des Venus-Durchgangs herausgegeben, six volumes, Reichsdruckerei, Berlin.
2. Becker, E. 1895, Catalog von 9208 Sternen zwischen $20 \circ 0'$ und $25^\circ 10'$ nördlicher Declination für das Äquinoktium 1875 nach Zonen-Beobachtungen am Pistor und Martins'schen Meridiankreise der Königlichen Sternwarte zu Berlin in den Jahren 1879 bis 1883 (Catalog der Astronomischen Gesellschaft. Erste Abth.; No. 10.), Astronomische Gesellschaft in Kommission bei Wilhelm Engelmann, Leipzig.
3. Becker, E. 1899, Mikrometer und Mikrometermessungen, in *Handwörterbuch der Astronomie* **IIIa**, Ed. W. Valentiner, Trewendt, Breslau.
4. Becker, E. 1906, Katalog von 8204 Sternen zwischen $1 \circ 42'$ und $6^\circ 10'$ südlicher Deklination 1855 für das Äquinoktium 1900 nebst einmalig bestimmten Örtern von weiteren 107 Sternen nach Zonen-Beobachtungen von J. Halm, M. Zwink, W.F. Wislicenus (†), E. Becker, B. Wanach u.a. am Repsoldschen Meridiankreise der Kaiserlichen Universitäts-Sternwarte zu Straßburg in den Jahren 1888 bis 1893, 1895 und 1903 bis 1905, bearb. von E. Becker (Katalog der Astronomischen Gesellschaft.

Zweite Abt.; No. 1.), Astronomische Gesellschaft in Kommission bei Wilhelm Engelmann, Leipzig.
5. Becker, E. 1912, Relative Bestimmungen der Intensität der Schwerkraft auf fünfundvierzig Stationen von Elsass und Lothringen nach den Beobachtungen von E. Becker, J. Bürgin, L. Carnera, P. Gast, E. Jost, H. Kobold, K. Schiller, G. Braunsche Hofdruckerei, Karlsruhe, 150 pp. + 1 map.
6. Becker, E. & Valentiner, W. 1906, Bestimmung der Längendifferenz zwischen der Großh. Sternwarte (Astronom. Institut) bei Heidelberg und der Kaiserl. Universitäts-Sternwarte in Straßburg i.E. im Jahre 1903 nach den Beobachtungen von L. Carnera, L. Courvoisier und W. Valentiner, G. Braunsche Hofdruckerei, Karlsruhe, 79 pp.
7. Bethge, K. & Klein, H. 1989, Physiker und Astronomen in Frankfurt, Alfred Metzner Verlag, Frankfurt-Neuwied.
8. Charlier, C.V.L. & Engström, F. 1904, Porträtgallerie der Astronomischen Gesellschaft, Hasse W. Tullberg, Stockholm. [PGAG 1904]
9. Cohn, B. 1918, Der Almanach Perpetuum des Abraham Zacuto. Ein Beitrag zur Geschichte der Astronomie im Mittelalter, *Schriften der Wissenschaftlichen Gesellschaft in Straßburg*, 32. Heft, Karl J. Trübner, Straßburg, 48 pp.
10. Craig, J.E. 1984, Scholarship and Nation Building. The Universities of Strasbourg and Alsatian Society, 1870-1939, Univ. Chicago Press, Chicago & London.
11. Cremer, M. 2001; Seismik zu Beginn des 20. Jahrhunderts – Internationalität und Disziplinbildung, *Berliner Beiträge zur Geschichte der Naturwissenschaften und der Technik* **28**, ERS-Verlag, Berlin.
12. Duerbeck, H.W. 2005a, Walter F. Wislicenus and Modern Astronomical Bibliography, this volume.
13. Duerbeck, H.W. 2005b, Strasbourg Observatory in World War II, this volume.
14. Duerbeck, H.W. & Seitter, W.C. 2005, The nebular research of Carl Wirtz, this volume.
15. Esclangon, E. 1926, La nouvelle organisation de l'observatoire, *Annales de l'Observatoire de Strasbourg*, Tome I, p. 1, Gauthier-Villars et Cie, Paris.
16. Foerster, W. 1911, Lebenserinnerungen und Lebenshoffnungen (1832 bis 1910), Georg Reimer, Berlin.
17. Gerland, G. 1906, Zirkular des Zentralbureaus der Internationalen seismischen Assoziation, *Astron. Nachr.* **172**, 63.
18. Hausmann, S. 1897, Die Kaiser-Wilhelms-Universität Strassburg – Ihre Entwicklung und ihre Bauten, Verlag W. Heinrich, Strassburg i.E., [see especially: Baugeschichte, p. 78; Kosten der Neubauten, p. 102; and Das astronomische Institut und die Sternwarte (by E. Becker), p. 145]
19. Ilgauds, H.J. & Münzel, G. 1995, Die Leipziger Universitätssternwarten auf der Pleißenburg und im Johannistal, Sax-Verlag, Beucha.
20. Kertz, W. 1999, Geschichte der Geophysik, Georg Olms Verlag, Hildesheim, Zürich, New York.
21. Kobold, H.A. 2004, Blätter der Erinnerung/Leaves of Memory, *J. Astron. Data* **10**.
22. Kohlschütter, A. 1948, Astrometrie (p. 75), in *Astronomie, Astrophysik und Kosmogonie*, Ed. P. ten Bruggencate, Dieterich'sche Verlagsbuchhandlung, Wiesbaden.
23. Rebeur-Paschwitz, E. von 1889, The earthquake of Tokyo, April 18, 1889, *Nature* **40**, 294-295.
24. Rebeur-Paschwitz, E. von 1892, Beobachtungen kleiner Erderschütterungen am selbstregistrierenden Horizontalpendel auf den Sternwarten zu Strassburg und Nicolaiew 1892, *Astron. Nachr.* **132**, 113.
25. Rebeur-Paschwitz, E. von 1894, Erdbeben in Beludschistan 20. Dec. 1892, *Astron. Nachr.* **135**, 335.
26. Rebeur-Paschwitz, E. von 1895a, Horizontalpendelbeobachtungen auf der Kaiserlichen Universitäts-Sternwarte zu Strassburg 1892-1894, *Beiträge zur Geophysik* **2**, 211-536.

27. Rebeur-Paschwitz, E. von 1895b, Vorschläge zur Errichtung eines internationalen Systems von Erdbeben-Stationen, *Beiträge zur Geophysik* **2**, 773-782.
28. Schiller, K. 1937, Die Beobachtungen der Polarissima BD +89°1 am Strassburger Meridiankreis, *Berichte der math.-phys. Kl. Sächs. Akad. Wiss. Leipzig*, 89. Bd.
29. Schramm, J. 1996, Sterne über Hamburg, Kultur- und Geschichtskontor, Hamburg.
30. Schur, W. 1884, Die Sternwarte, in *Festschrift zur Einweihung der Neubauten der Kaiser-Wilhelms-Universität Strassburg*, [Ed. Anonymous], C.F. Schmidt's Universitäts-Buchhandlung, Friedrich Bull, Straßburg, p. 79.
31. Schweitzer, J. 2003, Chapter 79.24 Germany/German National Report/Part A. Early German Contributions to Modern Seismology, in *International Handbook of Earthquake and Engineering Seismology* **81B**.
32. Schweitzer, J. 2004, Ernst von Rebeur-Paschwitz, the Father of Teleseisms, Talk presented at the 29th General Assembly of the European Seismological Commission, Potsdam, 12-17 September 2004.
33. Spamer, O. 1881, Illustriertes Konversations-Lexikon der Gegenwart, Spamer, Leipzig.
34. Tass, A. 1931, Portraitgallerie der Astronomischen Gesellschaft, Königliche Ungarische Universitätsdruckerei, Budapest. [PGAG 1931]
35. Winnecke, F.A.T. 1856, De stella η Coronae borealis duplici, Dissertatio astronomica inauguralis, Gustav Schade, Berlin.
36. Winnecke, F.A.T. 1875, Ueber die auf der Universitäts-Sternwarte zu Strassburg begonnenen Beobachtungen der Nebelflecke, *Vierteljahrsschrift Astron. Ges.* **10**, 297-304.
37. Wolfschmidt, G. 2005, Strasbourg Observatory in German Periods, this volume.

Appendix: The Imperial Strasbourg Observatory - A Collection of Published Sources

THE UNIVERSITY OBSERVATORY ANNALS

Five volumes of Annals of the Imperial University Observatory in Strasbourg were printed and published by G. Braun'sche Hofbuchhandlung in Karlsruhe between 1896 and 1926. Details about the contents are found in the chapter on Strasbourg Observatory Institutional Publications by Ph. Vonflie and A. Heck.

ANNUAL REPORTS OF STRASBOURG OBSERVATORY

Annual reports of the Imperial University Observatory in Strasbourg appeared in the following publications:

Annual reports of German observatories are usually published in the *Vierteljahrsschrift der Astronomischen Gesellschaft*, the predecessor of the present *Mitteilungen der Astronomischen Gesellschaft*. This custom was followed by Winnecke, Kobold, Becker, and Bauschinger (as well as Hellerich during the Nazi rule over Alsace). Only Schur published reports, covering longer time intervals, in the *Astronomische Nachrichten*.

In addition, the two reports on the edition of the *Astronomischer Jahresbericht* by Wislicenus are included.

1875 (A. Winnecke) "Ueber die auf der Universitäts-Sternwarte zu Strassburg begonnenen Beobachtungen der Nebelflecke", p. 297-304, VJS 10 (1875)
1876 (A. Winnecke) p. 81-86 in VJS 12 (1877)
1877 (A. Winnecke) p. 166-171 in VJS 13 (1878)
1878 (A. Winnecke) p. 167-171 in VJS 14 (1879)
1879 (A. Winnecke) p. 150-154 in VJS 15 (1880)
1880 (A. Winnecke) p. 253-255 in VJS 16 (1881)
1884 (W. Schur) "Beobachtungen und Untersuchungen am Repsold'schen Meridiankreise der Strassburger Sternwarte", AN 109, 129
1885 (W. Schur) "Bericht über die an den Instrumenten der Strassburger Sternwarte in den letzten Jahren ausgeführten Untersuchungen und Beobachtungen", AN 112, 161-176
1886 (W. Schur) "Fernerer Bericht über die Thätigkeit der Strassburger Sternwarte", AN 114, 401-404
1886 (H. Kobold) p. 155-160 in VJS 22 (1887)
1887 (E. Becker) p. 138-141 in VJS 23 (1888)
1888 (E. Becker) p. 154-167 in VJS 24 (1889)
1889 (E. Becker) p. 155-164 in VJS 25 (1890)
1890 (E. Becker) p. 158-165 in VJS 26 (1891)
1891 (E. Becker) p. 155-161 in VJS 27 (1892)
1892 (E. Becker) p. 225-235 in VJS 28 (1893)
1893 (E. Becker) p. 163-170 in VJS 29 (1894)
1894 (E. Becker) p. 188-196 in VJS 30 (1895)
1895 (E. Becker) p. 148-151 in VJS 31 (1896)
1896 (E. Becker) p. 120-124 in VJS 32 (1897)
1897 (E. Becker) p. 152-157 in VJS 33 (1898)
1898 (E. Becker) p. 167-173 in VJS 34 (1899)
1899 (E. Becker) p. 178-182 in VJS 35 (1900)
1900 (E. Becker) p. 149-154 in VJS 36 (1901)
1901 (E. Becker) p. 162-166 in VJS 37 (1902)
1902 (E. Becker) p. 157-160 in VJS 38 (1903)
1899-1902 (W.F. Wislicenus) p. 161-162 in VJS 38 (1903)
1903 (E. Becker) p. 138-143 in VJS 39 (1904)
1904 (E. Becker) p. 183-186 in VJS 40 (1905)
1903-1904 (W.F. Wislicenus) p. 187-188 in VJS 40 (1905)
1905 (E. Becker) p. 200-204 in VJS 41 (1906)
1906 (E. Becker) p. 212-215 in VJS 42 (1907)
1907 (E. Becker) p. 257-263 in VJS 43 (1908)
1908 (E. Becker) p. 277-285 in VJS 44 (1909)
1909 (J. Bauschinger) p. 263-267 in VJS 45 (1910)
1910 (J. Bauschinger) p. 189-192 in VJS 46 (1911)
1911 (J. Bauschinger) p. 159-162 in VJS 47 (1912)
1912 (J. Bauschinger) p. 149-152 in VJS 48 (1913)
1913 (J. Bauschinger) p. 210-214 in VJS 49 (1914)
1914 (J. Bauschinger) p. 134-138 in VJS 50 (1915)
1915 (J. Bauschinger) p. 143-146 in VJS 51 (1916)
1916 (J. Bauschinger) p. 250-253 in VJS 52 (1917)
1917 (J. Bauschinger) p. 225-226 in VJS 53 (1918)
1941 (J. Hellerich) p. 215-216 in VJS 77 (1942)
1942 (J. Hellerich) p. 181 in VJS 78 (1943)

THE LIST OF PHD THESES

The following list contains the titles of the theses, arranged chronologically. Those marked by [Aut] are in the possession of the author, or where inspected by him, those marked as [AIUB] are found in the library of the Astronomical Institutes of Bonn University, those marked as [SBB] are found in the Staatsbibliothek Berlin, and those by [BVB/Regensburg] in the Regensburg University Library. In the last two cases, titles and sizes were taken from the catalog entries.

Elkin, William, 1880: Über die Parallaxe von α Centauri, Inauguraldissertation der mathematischen und naturwissenschaftlichen Facultät der Kaiser-Wilhelms-Universität Strassburg zur Erlangung der Doctorwürde vorgelegt, Karlsruhe. Druck der G. Braun'schen Hofbuchdruckerei, 44 S. ([Aut])

Hartwig, Ernst, 1880: Beitrag zur Bestimmung der physischen Libration des Mondes aus Beobachtungen am Strassburger Heliometer, Inaugural-Dissertation der mathematischen und naturwissenschaftlichen Facultät der Kaiser-Wilhelms-Universität Strassburg zur Erlangung der Doctorwürde vorgelegt, Karlsruhe. Druck der G. Braun'schen Hofbuchdruckerei. 1880, IV + 51 S. + Lebenslauf (1880 Jul 28) ([Aut])

Küstner, Friedrich, 1880: Bestimmungen des Monddurchmessers aus neun Plejadenbedeckungen des Zeitraumes 1839 bis 1876: mit gleichzeitiger Ermittlung der Oerter des Mondes, Halle, Blochmann [Dr.], 1880, Nova Acta Leopoldina 41,1,5 (S. 256-364), 112 S., Strassburg, Univ., Diss. 1880, [BVB/Regensburg]

Wislicenus, Walter F., 1886: Beiträge zur Bestimmung der Rotationszeit des Planeten Mars, Inaugural-Dissertation der mathematischen und naturwissenschaftlichen Facultät der Kaiser-Wilhelms-Universität Strassburg zur Erlangung der Doctorwürde vorgelegt, Karlsruhe. Druck der G. Braun'schen Hofbuchdruckerei. 1886, 71 S. + 1 Taf. + Lebenslauf (Dez. 1885), Strassburg, Univ., Diss. 1886, [Aut]

Ristenpart, Friedrich, 1992: Untersuchungen über die Constante der Präcession und die Bewegung der Sonne im Fixsternsysteme, 96 S., graph. Darst., Strassburg, Univ., Diss. 1892, [BVB/Regensburg]

Cohn, Berthold, 1897: Über die Gauss'sche Methode aus den Beobachtungen dreier gleichen Sternhöhen die Höhe, Zeit und Polhöhe zu finden und praktische Hilfsmittel zu ihrer Anwendung, Strassburg i.E.: J. Singer, Buchh., 1897, 1 Bl., 70 S., 4 Taf., 4-to, Strassburg, Math.-naturw. Fak., Inaug.-Diss. von 1897, [AIUB]

Peck, Henry Allen, 1897: The equinox and obliquity of the ecliptic for 1885.0 and the constant of the lunar equation: from observations made upon the Repsold meridian circle of the Strasbourg Observatory, 1883-1888, Lynn, Mass., USA: the Nichols Press, 1897, 45 S., 1 Bl., 4-to, Strassburg, Math.-naturw. Fak., Inaug.-Diss. von 1897, [SBB]

Courvoisier, Leopold, 1900: Untersuchung über die absolute Polhöhe von Straßburg i.E., Heidelberg: Univ.-Buchdr. u. Verl. 1901, Heidelberg: Ph. Wiese, 1901, 131 S., 4-to, Strassburg, Math.-naturw. Fak., Inaug.-Diss. von 1900, [SBB]

Rosenberg, Hans [Oswald], 1905: Der Veränderliche χ Cygni, Halle a.S.: Ehrhardt Karras, 1906, (p. 121 to p. 258, 22 tables: Tab. III - XXIV.), 4-to, (from: Nova Acta, Abhandlungen der kaiserlichen Leopoldinisch-Carolinischen deutschen Akademie der Naturforscher, Bd. 85), Strassburg, Math.-naturw. Fak., Inaug.-Diss. von 1905 (Ref.: Becker), [Aut]

Plate, Heinrich, 1910: Punktausgleichung und Fehlerbestimmung nach graphischen Methoden in ihrer Anwendung auf Ortsbestimmung durch Standlinien, Strassburg i.E.: Müh, 1910, 47 S, 3 Bl., 8-vo, Strassburg, Math.-naturw. Dissertation vom 21 Dez 1909 (Ref.: Bauschinger), [SBB]

Redlich, Erich, 1911: Über die Bahn des Kometen 1886 I, Kiel: Oheim [Schaid?] 1911, 54 S, 4-to, (Aus: Astronomische Nachrichten Bd. 187, Nr. 4477-78), Strassburg, Math.-

naturw. Dissertation vom 21 Dez 1909 (Ref.: Bauschinger), [SBB]

Semenow, Leonid [Semenov, Leonid Ivanovic], 1911: Parallaxenbestimmung einer Gruppe von Zenitsternen in Pulkowo, St.-Petersbourg, Imp. "Slovo", 1911, 44 S., 1 Taf., 2-to, (From: Publications de l'Observatoire central Nicolas, Sér. 2, Vol. 18), Strassburg, Math.-naturw. Dissertation vom 12 Nov 1911 (Ref.: Bauschinger), [SBB]

Messow, Benno, 1913: Die beiden Sternhaufen im Perseus N.G.C. 869 und 884: Positionen und Helligkeiten nach photographischer Ausmessung, Hamburg 1913, (Lütcke & Wulff), 1913, 69 S., 4-to, Strassburg, Math.-naturw. Fak., Dissertation vom 17 Feb 1913 (Ref.: Bauschinger), [also: Astronomische Abhandlungen der Hamburger Sternwarte in Bergedorf, Bd. 2,2; and: Jahrbuch der Hamburger Wissenschaftlichen Anstalten, Band 30, Beiheft 4], [Aut]

Tschetschott, Anton, 1913: Untersuchungen über die kosmische Refraktion und über absolute Fixsternparallaxen. Angestellt an dem Strassburger Altazimut in den Jahren 1910-1911, Inaugural-Dissertation der mathematischen und naturwissenschaftlichen Fakultät der Kaiser-Wilhelms-Universität Strassburg zur Erlangung der Doktorwürde vorgelegt, Strassburg i.E., Elsass-Lothringische Druckerei, Abteilung Müh & Cie., Kinderspielgasse 20, 1913, 113 S. + 1 S. Lebenslauf, 4-to, [Aut]

Matzdorff, Martin, 1914: Berechnung des Mondradius aus Bedeckungen von Sternen erster Größe während der Jahre 1831 bis 1911, Karlsruhe i.B.: Braun, 1914, 60 S., 4-to, Strassburg, Math.-naturwiss. Dissertation v. 1914 (15. Dez. 1913), (Ref.: Bauschinger), [SBB]

Schumacher, Carl [Heinrich Josef], 1917: Untersuchungen über die ptolemäische Theorie der unteren Planeten Münster: Aschendorff, 1917, 78 S., Ill., 8-vo, Strassburg, Dissertation 1917 (9 Dez 1916) (Ref.: Bauschinger), [SBB]

STRASBOURG OBSERVATORY IN WORLD WAR II

H.W. DUERBECK
Vrije Universiteit Brussel
Pleinlaan 2
B-1050 Brussel, Belgium
hduerbec@vub.ac.be

Abstract. During World War II, the *Reichsuniversität Strassburg* was installed by the German authorities and Johannes Hellerich (1888-1963) was appointed director of the Observatory. A review of his life and his astronomical career as an assistant and professor in Kiel, Hamburg, Strasbourg, and Muenster is given. His activity in Strasbourg from mid-1941 to mid-1943 was focussed on bringing the Observatory into working operations, and on carrying out the monitoring of solar radiation and atmospheric extinction. After being drafted to the army and spending some time as a prisoner of war, he returned to Hamburg to complete a review on variable-star research in wartime Germany. He was called to Muenster University in 1947 to teach astronomy, and, from 1949 onward, to serve as director of the small Astronomical Institute till his retirement in 1954.

1. Johannes Hellerich – A life

In the short time the University of Strasbourg was run under German administration, it was Johannes Hellerich who led the Observatory.

Hermann Hinrich Peter Johannes Hellerich was born in Hamburg on 11 February 1888 as the son of the ship captain Johannes Hellerich. After finishing high school in Hamburg, he studied astronomy, mathematics and physics for two years in Jena and for four years in Berlin, where he was promoted in 1913 with a dissertation on δ Cephei stars (Hellerich 1913). It had been suggested by Paul Guthnick while Karl Hermann Struve and Fritz Cohn acted as referees. Hellerich summarized the photometric and spectroscopic observations, and interpreted them in the line of the – at that time accepted – binary hypothesis, with a sophisticated brightness distribution on the surface of the primary component. A year later, Shapley (1914)

A. Heck (ed.), The Multinational History of Strasbourg Astronomical Observatory, 123–132.

Figure 1. Walter Baade and Johannes Hellerich (right) in Hamburg, on the occasion of a lecture of Baade at the observatory on 14 February 1936. (Archive of Hamburger Sternwarte)

showed convincingly that the binary hypothesis should be abandoned and be replaced by the assumption of stellar pulsation.

In the following $1\frac{1}{2}$ years, he was non-budgetary [außerplanmäßiger] assistant in Bonn and was busy with meridian circle reductions, before he entered military service in the navy. Shortly before his release, World War I broke out, and he had to serve during the wartime years. After the war, he was offered the post of assistant at Kiel Observatory. He started work there on 1 May 1919, one month later than Carl Wirtz who had been offered the position of professor after his expulsion from Strasbourg. For the following years, both astronomers were mainly busy with meridian circle observations and reductions, while also pursuing their own research interests – Wirtz his statistical investigations of stars and nebulae, Hellerich his work on eclipsing binaries and variable stars.

In the early Kiel years, two papers by Wirtz and Hellerich dealt with cometary brightness observations (Hellerich & Wirtz 1920, 1921), but it appears that their observations – made with binoculars – were carried out independently and published under the same heading. Hellerich also wrote chapters on observation and calculation of binary stars for the advanced amateur book *Hevelius* (Hellerich 1922), as well as for the more prestigious *Lehrbuch der Physik* of *Müller-Pouillet* (Hellerich 1928). He obtained his *habilitation* in 1924 on the topic of eclipsing binaries (Hellerich 1925), and became a *nichtbeamteter* professor (*i.e.* appointed on a non-permanent basis) in 1928. In 1929 he became observer [Observator] and *nichtbeamteter* professor at Hamburg Observatory. There he met Walter Baade (Fig. 1), who obviously shared his interest in Cepheids. Hellerich continued his work on variable and eclipsing stars, and also contributed two major articles on visual photometry and colorimetry for the *Handbuch der Experimentalphysik* (Hellerich 1937).

Obviously primarily to foster his career, he applied for membership in the National Socialistic German Workers' Party (NSDAP) by 30 September 1937, when still being an observer in Hamburg. After the beginning of World War II, he was drafted again and served in the Wilhelmshaven wharf. Declared u.k. (*unabkömmlich, i.e.* reserved) in Summer 1941, he was nominated, on 28 August 1941, full professor for astronomy and astrophysics and director of the *Sternwarte der Reichsuniversität Strassburg* – and his party membership was certainly helpful for obtaining this job, although not necessarily a sufficient condition! However, his Strasbourg career was cut short because he was called up again to the army.

After the German defeat, he was interned by the French, spending some time in a camp at Saint-Sulpice-sur-Tarn near Albi in Southern France. A notebook from this time still exists in his estate, with entries about his life, illness, and estimates of variable star magnitudes. In February 1946, he was released and returned to his former residence, Hamburg. There, by invitation of Otto Heckmann, director of Hamburg Observatory, he served as a scientific collaborator, and completed a survey of variable-star research in wartime Germany (Hellerich 1948), which forms part of the so-called Field Information Agencies Technical (FIAT) review in astronomical research during the years 1939-1946. He looked for positions in astronomy in Hamburg, Kiel, Freiburg, and Berlin, and, in 1947, he obtained a guest professorship in Muenster. In March 1949, he took over the newly formed extraordinary professorship and the directorship of the Astronomical Institute of Muenster University. Major parts of the university and the city were in ruins, and there was basically no infrastructure nor budget. Rooms for the institute were made available in the rural castle Buldern. Hellerich retired in 1956, and died on 30 May 1963 (Strassl 1964).

2. German Astronomical Activity During WW II in Strasbourg

There is little written evidence for activity at Strasbourg Observatory during the German occupation. Two relevant documents were found in the Bundesarchiv Berlin (Sign. R76 IV).

In a letter (Fig. 2) written on 22 September 1941 to the Rector of the University, the geographer Georg Niemeier, Dean of the Faculty of Natural Sciences, lists 19 professors "to be appointed for the Winter term 1942 according to information by the Ministry of Science and Education" (Reichsministerium für Wissenschaft, Erziehung und Volksbildung, abbreviated REM). Among them is Johannes Hellerich, as "full [ordentlicher] professor for astronomy." In the annual report of 1941, Hellerich wrote "by 1 July 1941 I provisionally was charged with the direction of the observatory, on 1 August 1941 the definitive appointment followed."

The filling of the position of the observer turned out to be problematic. Another letter from the Ministry to the Rector in Strasbourg, dated 30 August 1941, reads: "Docent Dr. Walther [sic], who, according to the proposal of Professor Dr. Hellerich has been suggested to fill the post of observer, will presumably be appointed to Cracow [Krakau]. I propose, in his place, the docent Dr. Oswald Wachtl in Bonn to fill the beforementioned post, and I request a report whether there exist doubt against the person Dr. Wachtl. On behalf, [signature]."

Kurt Walter (1905-1992), who had studied mathematics and physics in Stuttgart, and astronomy in Koenigsberg (today Kaliningrad), got his PhD and habilitation in Koenigsberg in 1928 and 1932, respectively. Besides his obligations (meridian circle observations, time service, calculations), his main, life-long interest was in the field of close binary stars. Having been employed as a calculator in Koenigsberg, he moved to Potsdam Observatory as a scientific employee in early 1937. He became non-budgetary [außerplanmäßiger] professor at Berlin University in November 1941. Since October 1941, he had been provisionally charged with the direction of Cracow observatory, and, starting 1942, had indeed became director of the [Polish] Observatories in the Generalgouvernement, and thus supervisor of all Polish astronomers. See his annual report (Walter 1943), as well as his certainly biased, but nevertheless informative *Memories of a Participating Astronomer* (Walter 1987). After the war, he first made his living as a farm-hand, then served as a librarian at the state library in Stuttgart, before becoming a professor of astronomy at Tuebingen University (1962-1972). An obituary was written by Mauder (1993).

The other prospective candidate, Oswald Wachtl (1906-1942), had studied and received his PhD in Vienna. In 1930 he had joined Bonn Observatory, where he was mainly occupied with plate measurements of the Astronomische Gesellschaft star catalog (AGK2). He obtained his habilitation in 1937 (Wachtl 1937). In World War II, he served as a lieutenant in the German Air Force and was killed on 8 March 1942 during a flight at the Channel coast (Sticker 1942).

23. SEP. 1941

Der Dekan
der Naturwissenschaftlichen Fakultät
der Universität Straßburg

Straßburg, den 22.September 1941.
Universitätsstraße 11
Fernruf 205.97

Ihr Zeichen

Unser Zeichen 41-9-89 Nie./JC.

An den
Herrn R e k t o r
der Universität Straßburg

Mit der Berufung folgender Kollegen der Naturwissenschaftlichen Fakultät zum Wintersemester 42 kann laut Auskunft des REM gerechnet werden:

Beischer, Dietrich, a.o.Prof. für Anorganische Chemie
Dehm, Richard, a.o.Prof. für Paläontologie
Drescher-Kaden, Friedr.Karl, o.Prof.f. Mineralogie
Firbas, F., o.Prof. für Botanik
Grewe, Rudolf, a.o.Prof. für Organische Chemie
Hegemann, Friedrich, a.o.Prof. für Geochemie
Hellerich, Johannes, o.Prof. für Astronomie
Hiedemann, Egon, o.Prof. für Technische Physik
Holleck, Ludwig, Doz. für Phys.Chemie
Jander, Wilhelm, o.Prof. für Anorganische Chemie
Jung, Karl, o.Prof.f. Geophysik
Knapp, Edgar, o.Prof. für Phytogenetik
Niemeier, Georg, o.Prof. für Geographie
Noddack, Walter Karl Friedr., o.Prof.f.Phys.Chemie
Petersson, Hans, o.Prof. für Mathematik
Schlemmer, Ferdinand, o.Prof. für Pharmazeutik u.Pharmakognosie
Weber, Hermann, o.Prof. für Zoologie
Wilckens, Otto, o.Prof. für Geologie
Zeile, Karl, o.Prof. für Organische Chemie

Figure 2. Letter announcing the appointment of J. Hellerich to become professor of astronomy at the Universität Strassburg. Conspicuously absent are the professors for experimental and theoretical physics. (Sign. R76 IV; copy from the Bundesarchiv Berlin)

Thus, the persons selected to assist Hellerich in his work never made it to Strasbourg. In the two annual reports 1941/1942 for Strasbourg Observatory (Hellerich 1942, 1943), he mentions the following employees: the scientific assistant J. Huß (who was also active in pre- and post-war times in Strasbourg), the mechanic R. Schneider, the technical clerk A. Ehrhardt, the gardener/housekeeper L. Wetzel, the computer/librarian Miss E. Rühl, the scientific helper stud. astr. K. Schrey (15 June to 30 September 1942) who was replaced by Miss A. Breitung on 1 October 1942.

Hellerich also mentions that the main building of the observatory first served to house the rector's office, which moved out of the building by 1 September 1942. Before, the observatory's office spaces were housed in the laboratory building which had been built in 1935.

Because of the evacuation of Strasbourg in 1939, the meridian circle, the main clocks and the optics of the larger instruments had been moved to Paris or Bordeaux, while the library, the minor instruments and the workshop tools had gone to Clermont-Ferrand. The items were returned to Strasbourg in late 1941. Repairs of the electric motion of the refractor dome, as well as electric supply lines in the refractor and meridian buildings were started in the second half of 1941. While the dome repair was finished in 1942, the supply line repairs were still not completed.

There was no indication of any type of scientific activity in Hellerich's Strasbourg reports. Strassl (1964) mentions that Hellerich gave classes in Strasbourg as well as in nearby Freiburg im Breisgau since astronomy was not represented at that university. The observing activity was confined to radiation measurements of the Sun to determine the extinction factor of the Strasbourg region; several files of observations (from 18 February to 24 August 1943) and reductions (using his own and earlier French material) survive in his written estate kept at the University Archive in Muenster. Hellerich also evaluated the scientific estate of the long-time editor of the *Astronomische Nachrichten*, Hermann Kobold, and wrote a report (Fig. 3) to the Ministry of Education (REM) on 2 November 1942.

This is also of interest for the history of astronomy at Strasbourg since a long-time project of Kobold concerned the reduction of measures of the solar diameter made with the Strasbourg heliometer from 1877 to 1885, and from 1886 to 1899; the latter series was mostly observed by Kobold himself. Reductions had been delayed for a long time because of Kobold's editorship (1925-1938) and then because of his progressing blindness. In 1938, Kobold started to obtain funds from the REM to pay an assistant, and the project made some progress; yearly reports and applications for funds are found in the Kiel Acta of the Bundesarchiv. After Kobold's death in June 1942, Hellerich in Strasbourg was obviously the right person to decide on the further procedure and he opted for a continuation of the reductions by the assistant in Kiel. He also indicated that at a later stage a Strasbourg assistant would be able to take over the project, and he agreed to finally discuss and publish the results himself.

The material is not kept in Hellerich's estate in Muenster; according to his early post-war correspondence, it was left behind in Strasbourg. While it seems that in 1946, Huß brought some material to Germany, it is not clear whether the Kobold observations were among it, maybe it was the data on radiation measurements. It is not impossible that it is still shelved somewhere at Strasbourg observatory; maybe parts of it can also be found in Kobold's estate which is kept in the Bibliothek of the Christian-Albrechts-Universität Kiel (Lorenzen-Schmidt 1979).

3. After the War

An interesting circular letter is found in Hellerich's estate, written on 5 February 1946 by the former Dean of the Faculty of Natural Sciences, Georg Niemeier, where he describes his difficulties with the Ministry of Education and party authorities to appoint professors. Let me quote a few statements:

I can explicitly say that no colleague was appointed to our Faculty because he was a party member, the same is valid for all habilitations.

As a proof for the accomplishment of this principle and for the struggle against the NS-bigwig business [NS-Bonzenwirtschaft] *the following events should me mentioned: The accomplishment of appointments of several colleagues who were not party members (*e.g. *Hegemann, Hiedemann, von Weizsäcker), and the rejection of "politically-proven" people (which had also been rejected by other German universities before because of their poor scientific qualities).*

From the first day of the deanship up to the collapse there was an ongoing struggle for the filling of the professorial chairs in physics. In this matter, the small group of the "German Physics"[1] *played a role, which should be instituted by force by the REM [...] and thus up to the collapse, the chair of experimental physics was not properly filled because of intrigues of the REM and the party office*[2]*. Colleague Finkelnburg was only an extraordinary professor and only administered the professorial chair of experimental physics, his "crime" was that he did not stand on the side of "German Physics", and tried to start discussions between different camps, in order to extinguish the hateful conflicts concerning certain facts of Einstein's theory of relativity that had been proven correct a long time ago.*

[1]On "German Physics" during Nazi times, see for instance Beyerchen (1977).

[2]Among the surviving Strasbourg papers in the Bundesarchiv, there is a handwritten letter by Philipp Lenard, 1905 Nobel Prize winner and author of the *Deutsche Physik* of 1936. In this letter to the minister, dated 7 May 1941, he suggests a certain experimental scientist with "proven national-socialistic thinking" for the position.

Sternwarte
der Reichsuniversität Strassburg

B.-Nr.848

Strassburg, den 2.November 1942
Universitätsstrasse 11
Postruf 71140

<u>Bericht</u>

betreffend Bücherei und wissenschaftlicher Nachlass
von Professor H. Kobold in Kiel.

<u>Vorgang:</u> Schreiben vom REM vom 24.7.1942 - W I Nr.1278 -

Ich habe im September 1942 die gewünschte Überprüfung des Nachlasses von Prof. Kobold vorgenommen und möchte im Einvernehmen mit Herrn Pastor Kobold in Preetz folgende Regelung vorschlagen.

1) <u>Bibliothek.</u> Neben dem Teil der Bibliothek, welcher seinen Erben verbleibt, sind noch eine Reihe von Veröffentlichungen ausländischer Sternwarten sowie Jahrgänge von periodischen Zeitschriften vorhanden, die Herr Pastor Kobold für eine Verteilung an deutsche Sternwarten zur Verfügung stellen will. Um einen Überblick über die Bedürfnisse der in Frage kommenden Sternwarten (Graz, Posen, Prag) zu gewinnen, habe ich Durchschriften der anliegenden Liste an diese Sternwarten mit der Bitte gesandt, mir bis zum 1.Dezember 1942 anzugeben, welche Bücher gewünscht werden. Nach Eingang der Listen werde ich einen Vorschlag zur Verteilung der Bücher zusenden. Die zurzeit noch in Kiel lagernden Bücher werden dann den Sternwarten auf deren Kosten zugesandt.

Für die unter b) genannten Reduktionsrechnungen des Teil
und für die unter c) angeführten erforderlichen Abschlussrechnu
stehen keine Geldmittel zur Bezahlung einer Hilfskraft zur Verf
Das zurzeit vorhandene Personal der Strassburger Sternwarte ist
den Aufbauarbeiten der Sternwarte (Aufstellung der Instrumente,
nung der Bibliothek usw.) voll beschäftigt und kann daher zur A
führung dieser umfangreichen Rechnungen nicht herangezogen werd
Ich bitte daher, mir für den Abschluss der Arbeiten, der voraus
lich in einem Jahre erfolgen wird, die Geldmittel in Höhe von
RM 2400 für das Rechnungsjahr 1943 zur Bezahlung einer Rechenh
zur Verfügung zu stellen.

Wegen des Druckes und der Veröffentlichung wird eine Ent
dung erst getroffen werden können, wenn sich der Umfang des Ma
skriptes übersehen lässt.

Hellerich

Figure 3. Fragments of a surviving letter of Hellerich written during his Strasbourg time, dealing with the Kobold estate (books and reductions of solar heliometer observations). (Acta Kiel, Sign. R4901, No. 14811, p. 401; copy from the Bundesarchiv Berlin)

Pressure was exerted by the REM by scaling down salaries of appointed colleagues, with the outspoken reason: "we will show you that we still have the say in Str., although you don't pay attention to our suggestions for appointments."

He finally mentions that before the re-opening of the university, the faculty had been asked to consider that each institution should start its lectures with a "political credo". Within minutes the faculty members had unanimously agreed that this was scientifically stupid, and that they would give a real service to Germany and the borderland Alsace, not by issuing political phrases, but by scientific status and efforts.

Returning to Hellerich's party membership, I should add some very friendly post-war correspondence is found from the "non-party-member Hiedemann" and the Austrian astronomer Kasimir Graff, who had suffered a lot during Nazi times. This again confirms that Hellerich was one of the many fellow-travellers ["Mitläufer"] during the Nazi era.

While Hellerich was quite active in publishing papers while he was working in Kiel and Hamburg, his publication activity essentially ceased in 1939. However, after the war, he took an active part in the re-established Astronomische Gesellschaft, gave colloquia and popular lectures in different places. A paper on "The Phase Shift of Lightcurves of δ Cephei Variables at Different Wavelengths", emulating the Baade-Wesselink-method, was submitted to the *Astronomische Nachrichten* Mid-1949. Hellerich, obviously having overlooked Wesselink's (1946) paper due to the poor library means available to him, received a negative response from the editor, Johannes Wempe, who drew his attention to Wesselink's papers, published in the *Bulletin of the Astronomical Institutes of the Netherlands* in 1946/47, which had not been accessible to Hellerich (Nachlaß Hellerich, file 178). This seems to have been Hellerich's last attempt to publish a scientific paper. Work on eclipsing binaries and the chronology of Maya astronomy that he pursued in Muenster remained fragmentary and unpublished. A lot of material by himself, and copies and collections from other observers, is kept in his written estate at the Muenster University Archive.

Sources and Acknowledgments

Documents are kept in the Universitäts-Archiv Münster (Nachlaß Hellerich, 179 files kept in 18 boxes), and in the Bundesarchiv (Berlin), (NSDAP-Zentralkartei; Sign. R4901, No. 14811 (Kiel); Sign. R76 IV (Strassburg)). The latter ones are sparse since most of the German documents concerning the university in WW II were destroyed *in a state of panic by university officials* before the town was returned to the allied forces. Other documents concerning employees were lent to the federal administrative authority and were reported missing since 1984.

I thank the Universitäts-Archiv Münster, especially Ms. Annett Chollowa, and the Bundesarchiv (Berlin) for granting access to documents, and the Universitätsbibliothek Kiel for information on Kobold's estate. I am especially grateful to Prof. Dr. H.-J. Wendker (Hamburg) for providing the photograph of Baade and Hellerich.

References

1. Beyerchen, A.D. 1977, Scientists under Hitler: Politics and the Physics Community in the Third Reich, Yale Univ. Press, New Haven, xii + 287 pp. (ISBN 0-30001-83-0)
2. Hellerich, J. 1913, Neue Bearbeitung der photometrischen und spektroskopischen Beobachtungen der Veränderlichen vom δ Cephei-Typus, Inaugural-Dissertation zur Erlangung der Doktorwürde genehmigt von der Philosophischen Fakultät der Friedrich-Wilhelms-Universität zu Berlin (privately published: Carl Georgi, Printer, Bonn 1913).
3. Hellerich, J. 1922, Beobachtung und Berechnung der Doppelsterne, in *Hevelius. Handbuch für Freunde der Astronomie und kosmischen Physik*, Ed. J. Plassmann, F. Dümmler, Berlin, p. 185.
4. Hellerich, J. 1925, Untersuchungen über Bedeckungsveränderliche, *Astron. Nachr.* **223**, 369.
5. Hellerich, J. 1928, Doppelsterne und Veränderliche, in *Müller-Pouillets Lehrbuch der Physik*, Bd. 5, 2. Hälfte, Ed. A. Kopff, F. Vieweg, Braunschweig, pp. 289-336.
6. Hellerich, J. 1937, Visuelle Photometrie *and* Die Methoden der Farbenmessung, in *Handbuch der Experimentalphysik, Bd. 26: Astrophysik.* Ed. B. Strömgren, Akademische Verlagsgesellschaft, Leipzig, pp. 565 & 649.
7. Hellerich, J. 1942, Strassburg (1941), *Vierteljahrsschrift Astron. Ges.* **77**, 215.
8. Hellerich, J. 1943, Strassburg (1942), *Vierteljahrsschrift Astron. Ges.* **78**, 181.
9. Hellerich, J. 1948, Veränderliche Sterne, in *Astronomie, Astrophysik und Kosmogonie (Naturforschung und Medizin in Deutschland 1939-1946, Band 20 = FIAT Review of German Science)*, Ed. P. ten Bruggencate, Dieterich'sche Verlagsbuchhandlung, Wiesbaden, p. 335.
10. Hellerich, J. & Wirtz, C. 1920, Helligkeitsschätzungen des Brorsen-Metcalfschen Kometen 1919b. *Astron. Nachr.* **212**, 181.
11. Hellerich, J. & Wirtz, C. 1921, Schätzungen der Totalhelligkeit des Kometen 1921a (Reid), *Astron. Nachr.* **214**, 359.
12. Lorenzen-Schmidt, K.-J. 1979, Nachlaß H. Kobold, typoscript, Universitätsbibliothek Kiel.
13. Mauder, H. 1993, Kurt Walter (obituary), *Mitteil. Astron. Ges.* **76**, 9.
14. Shapley, H. 1914, On the nature and cause of Cepheid variation, *Astrophys. J.* **40**, 448.
15. Sticker, B. 1942, O. Wachtl (obituary), *Himmelswelt* **52**, 42.
16. Strassl, H. 1964, Johannes Hellerich (obituary), *Mitteil. Astron. Ges.* **17**, 27.
17. Wachtl, O. 1937, Studien zur endlichen Fortpflanzungsgeschwindigkeit der Gravitation, *Astron. Nachr.* **262**, 353.
18. Walter, K. 1943, Jahresbericht der Sternwarten des Generalgouvernements für die Jahre 1940-1943, *Vierteljahrsschrift Astron. Ges.* **79**, 42.
19. Walter, K. 1987, Astronomy in Poland during the Second World War, *J. British Astron. Assoc.* **97**, 270.
20. Wesselink, A.J. 1946, The Observations of Brightness, Colour and Radial Velocity of Delta Cephei and the Pulsation Hypothesis, *Bull. Astron. Inst. Netherl.* **10**, 91.

STRASBOURG OBSERVATORY: A BREEDING PLACE FOR FRENCH ASTRONOMICAL INSTRUMENTATION IN THE 20TH CENTURY

S. DÉBARBAT
Observatoire de Paris
61, avenue de l'Observatoire
F-75014 Paris, France
debarbat@danof.obspm.fr

Abstract. Between the two World Wars, Strasbourg Observatory became an instrumentation breeding place for the equipment of French astronomical observatories. This was certainly due to the presence of specialists in various fields, from applied physics to astronomy. This was probably also linked to the fact that, in a unit of moderate size, the exchanges of views and ideas are not overdiluted.

1. Introduction

At the end of World War I (WWI), the French government put quite a number of brilliant scientists in various disciplines into university positions in Strasbourg. This was for instance the case for the physicist Gabriel Foëx or for René Thiry, a specialist in celestial mechanics. With its links to the university, the astronomical observatory was also part of that recognition. Astronomers found in Strasbourg a well-equipped observatory[1].

2. Strasbourg Observatory between Two Wars

At the outcome of the Franco-Prussian war, the astronomer put in charge of constructing Strasbourg Observatory came from Pulkovo, the prestigious observatory built after Wilhelm Struve had been called there to set up a place as well equipped as Dorpat (nowadays Tartu, Estonia) where he had worked so successfully.

[1]See the chapters by Duerbeck (2005) and Wolfschmidt (2005) in this volume.

A. Heck (ed.), The Multinational History of Strasbourg Astronomical Observatory, 133–151.

Thus August Winnecke arrived in Strasbourg in 1872. Through his many publications, he has had a truly international impact with papers in the *St Petersburg Acad. Bulletin*, in the *Monthly Notices of the Royal Astronomical Society*, in the *Astronomische Nahrichten* as well as in the *Comptes Rendus de l'Académie des Sciences de Paris.*

Winnecke's poor health led to his replacement in 1887 by Ernst Becker. Becker was leaving the directorship of the famous Gotha Observatory (on the Seeberg) created by the Duke of Saxe-Coburg-Gotha and that Franz Xaver Baron von Zach (1754-1832) had directed from 1786 on. Julius Bauschinger succeeded Becker in Strasbourg in 1909. Bauschinger had worked in Munich from 1878 to 1881, then in Berlin in 1881/1882 before coming back to Munich where he stayed at the Observatory from 1883 to 1895. In 1896, he was nominated Professor of Astronomy at the University of Berlin and, from 1901 onwards, Director of the Astronomisches Rechen Institut, responsible at the world level for the *Berliner Astronomiches Jahrbuch.* Bauschinger was expelled at the end of WWI.

Through those facts, it is clearly visible that well-studied choices dictated the nomination of Strasbourg Observatory directors. France, after the departure of Bauschinger, could only choose men who had proven to be talentuous enough to take over.

3. Esclangon, the Director Between the World Wars

Ernest Esclangon was born on 17 March 1876 in Mison (now in the Alpes-de-Haute-Provence *département*). His parents were cultivators who had become landowners.

Esclangon was first educated at the College of Manosque, then at the *Lycée* [high school] of Nice where his capabilities were first noticed. In 1895, he entered the *École Normale Supérieure (ENS)* from which he graduated second in 1898 with an *agrégation* [teaching qualification] in mathematical sciences.

The following year, Esclangon got a position as "Aide-Astronome"[2] at Bordeaux Observatory. In 1902, he was put in charge of lecturing rational mechanics at the Faculty of Sciences of Bordeaux University. In 1905, and simultaneously to his function of deputy professor for differential calculus and infinitesimal geometry, he was nominated as "Astronome-Adjoint" at Bordeaux Observatory. His career got then oriented towards the mathematical sides of astronomy.

[2]The sequence (bottom-up) of the positions that astronomers could then hold in France were: *Assistant d'Observatoire* [Observatory Assistant], *Aide-Astronome* [Assistant Astronomer (confusing with previous one, we know)], *Astronome-Adjoint* [Deputy Astronomer], *Astronome Titulaire* [Titular Astronomer]. (Ed.)

Then World War I (WWI) broke out and he was assigned to the commission for naval artillery in the service of ballistics and sound ranging. That assignment followed a presentation he had made on the principle of localization through sound, as early as September 1914, to the geographical service of the army that had withdrawn to Bordeaux. Other researchers had worked on the same subject, of primary importance at that time, but Esclangon seems to have been the first one to bring it to light.

At the beginning of 1915, Esclangon was in Gâvres (Ille-et-Vilaine) where the naval artillery was located and where he worked on means to apply at the front a method based on the record of gunnery-generated waves.

At the end of the WWI, Esclangon was nominated Director of Strasbourg Observatory and quasi-simultaneously – as this will be the case for a long time – Professor of Astronomy at the Faculty of Sciences of that town. The destruction that had occurred in the Northern part of France required in all fields a cooperation between the most dynamical elements in those regions. At Strabourg Observatory, Esclangon benefitted from excellent collaborators during the ten years of his directorship (1919-1929).

Beyond his directorial and educational activities, Esclangon carried out studies on positional astronomy, on instrumentation, on chronometry in the line of his 1904 thesis on quasi-periodic functions. Trained initially in mathematics at the ENS, Esclangon has also been interested in applications of celestial mechanics and, in 1947, he had already envisaged the launch of an artificial satellite around our planet.

After investigating the possibilities to increase the precision of the Strasbourg meridian instrument, Esclangon studied in 1919 sets of gear wheels for replicating the relationship between the duration of the sidereal day and the mean solar day.

In 1921, he installed at Strasbourg Observatory two clocks providing the observers with the mean time through a synchronization with the sidereal one. The lower dial displayed the local sidereal time needed to know the time of meridian crossing for the celestial objects they wanted to observe. The upper dial provided Greewich mean time. Already since 1911, France was adhering to the system of time zones, the origin of which had been fixed at the Greenwich meridian as early as 1883-84. At Paris Observatory, Esclangon installed in 1934 a network of two-dial clocks in the observing outlets as well as in the astronomers' offices.

While, in Esclangon's era, the mean time was obtained from the sidereal time, today the time displayed – aligned on the coordinated universal time of physical origin (UTC) – uses the same set of gear wheels to provide the local sidereal time. When Henri Deslandres (1853-1948), who had succeeded in 1927 Benjamin Baillaud (1848-1934), left the directorship of Paris

Observatory in 1929, Esclangon in turn left Strasbourg for Paris where he took up the directorship on 1 November 1929. He became *ex officio* Director of the Bureau International de l'Heure (BIH) [International Time Bureau] and got interested in time issues, thinking especially of making time broadly available to the public.

In Strasbourg, there was already such a distribution by controlled signals emitted every minute. But, with the advent of talking movies in 1927, Esclangon considered a more practical solution as the usage of telephone was also spreading quickly. Two firms were contacted for realizing an almost permanent broadcast of time by spoken messages. The proposal by the Brillé company was retained and installed at Paris Observatory in 1932.

The first talking clock in the world was thus put into public service in 1933. The first day, some 140 000 phone calls attempted to get the time. Only 20 000 of them could be satisfied thanks to the twenty phone lines installed. But ten new lines had to be added, broadcasting the voice of Marcel Laporte who, under the name of Radiolo, was then a well-known radio speaker. In fact, the tapes were also carrying records by an engineer of the Brillé company, but because of the short duration of the messages, it was difficult to distinguish the two voices. Other talking-clock systems were subsequently set up, for instance in Great Britain in 1936.

In 1920 in Strasbourg, Esclangon had received the *Grand Prix des Sciences Mathématiques* of the French Academy of Sciences. That institution had already granted him prizes earlier in 1914 and 1916. From 1929 onwards, Esclangon was also a member of the Academy of Sciences (where he succeeded Pierre Puiseux) and a member of its commission on prizes. In 1932, he entered the Bureau des Longitudes (BdL), replacing Guillaume Bigourdan. Esclangon had become quite a personality recognized by his peers. And this was similarly the case at the international level.

Quickly involved in the creation of the International Astronomical Union (IAU) just after WWI, Esclangon made no mystery that he enjoyed travelling and opportunities to meet numerous colleagues. In 1928, he attended the IAU meeting in Leiden (The Netherlands). He then went to Cambridge (USA) in 1932.

The following IAU General Assembly (GA) took place in Paris in 1935 and Esclangon did not spare his energy to organize the event that included a reception at the Elysée Palace by Albert Lebrun (1871-1950), then President of the French Republic. At the end of the conference, Esclangon was elected IAU President for the period 1935-1938. He thus chaired the following GA in 1938 in Stockholm (Sweden).

The following year saw the outset of World War II. Esclangon organized the withdrawal of the time services to Bordeaux while this had been done

Figure 1. Double-dial clock (mean and sidereal times) installed by E. Esclangon at Strasbourg Observatory (Esclangon 1926). (© Strasbourg Astron. Obs.)

to Lyons in WWI. At the end of WWII, Esclangon was replaced by André Danjon as Director of Paris Observatory (see hereafter).

Beyond his scientific activities and the reorganization of Strasbourg Observatory, Esclangon's pedagogical qualities have to be remembered too – both in Strasbourg and in Paris – as often stressed by past students. His authority was also very much appreciated at the various conferences he attended and in the commissions where he worked.

Inquiring mind and and gracious man, he retired to Eyrenville, a hamlet of Dordogne, where he moved around on bicycle and where he died on 28

January 1954[3]. It is perhaps slightly unfortunate that his name remains very specifically associated to the creation of the talking clock. It is true though that the importance of time in everyday life, beyond its criticality for many research activities, only increased in the course of Esclangon's career in Strasbourg and later in Paris, as well as in the following decades till today.

4. Danjon: taking over in Strasbourg and Paris

André Danjon was born in Caen (Calvados) on 6 April 1890. This is why the *Journées "Systèmes de Référence Spatio-Temporels 1990"* [Space-Time Reference Systems 1990] became a 'Danjon Colloquium' (Capitaine & Débarbat 1990) with the participation of a number of his past students and collaborators. After attending the *Lycée Malherbe* in Caen, Danjon entered the *École Normale Supérieure (ENS)* in 1910. Four years later, he graduated as *Agrégé des Sciences Physiques.*

From the beginning of World War I (WWI), Danjon was mobilized, first in infantry, later at the service of sound ranging directed by E. Esclangon. At the end of WWI, Danjon was nominated professor at the military center preparing to the French *grandes écoles* [higher education schools] in Strasbourg. A few months later, he entered Strasbourg Astronomical Observatory, being immediately put in charge of teaching at the Faculty of Sciences.

Strasbourg Observatory was then well equipped, but virtually without personnel, and it was necessary to put it back to operations. Under the direction of Esclangon, Danjon was quickly associated to restarting the institution. He did not neglect other things such as his thesis in the field of astronomical photometry, the foundation of which dates back to Pierre Bouguer (1698-1758) and that had been largely explored by François Arago (1786-1852) in the 19th century. Thus, in his thesis presented in 1928 at Strasbourg Observatory and entitled *Recherches de Photométrie Astronomique* [Investigations in Astronomical Photometry], Danjon introduced his "cat-eye" photometer.

After studying observational conditions, he refined methods allowing to compare point sources. He also investigated the earthshine on the Moon. This was the first study of the kind, comparing the two lunar zones: the one illuminated by the Sun and the other one illuminated by the Earth only. Among other results, Danjon provided an average value for the Earth albedo.

In 1990 at the Danjon Colloquium, US astronomers presented an ongoing study of the Earth albedo (Weidman *et al.* 1990). Having built a cat-eye

[3]He has been buried in Mison.

photometer, they were aiming at studying the variation of the illumination produced by the Earth through long-term photometric measurements similar to those carried out by Danjon in Strasbourg from 1926 to 1935 and continued by J. Dubois at Bordeaux Observatory from 1940 to 1950. They were hoping to derive results on the evolution of the Earth atmosphere since the end of the 1920s, with special attention to the seasonal variations as Danjon did it in his own time.

As did Arago, Danjon also used a double-image photometer in order to obtain lightcurves for several typical variable stars. He observed also the lightcurves of Mercury and Venus. As to the double-image technique applied to double stars, Danjon got Paul Muller to study a novel micrometer while he himself used the double image through interferences for measuring close double stars. From all that activity based on the double-image technique, another instrument would later emerge, using the same optical principle: Danjon's prism astrolabe equipped with an impersonal micrometer (see *e.g.* Danjon 1958).

With his PhD secured in 1928, Danjon was nominated soon after (1929) as *Astronome-Adjoint.* He was then put in charge of lecturing upper-level astronomy at the Faculty of Sciences of Strasbourg University.

General Ferrié (1868-1932), who had launched in France the *Télégraphie Sans Fil (TSF)* [wireless telegraphy] at the Eiffel Tower, had developed an interest in astrophysics since the end of WWI. Under his stimulation, Danjon established, as early as 1923, an *Avant-projet d'organisation d'un observatoire d'astronomie physique* [Draft Project on Organizing an Astrophysical Observatory]. The envisaged program focussed on stellar astronomy, a field where, since a couple of decades, a number of discoveries had been made. The document also listed the necessary instruments and envisaged a laboratory, a workshop, a library and, of course, *ad hoc* personnel. The project was examined by a few advisers and took shape after some discussions. The search for a site was initiated in 1924, leading in 1932 to the selection of a location in the *Alpes de Haute-Provence* for what will become the *Observatoire de Haute-Provence (OHP)* near Saint-Michel, later renamed as Saint-Michel-l'Observatoire.

When Esclangon took over the directorship of Paris Observatory after the retirement of Henri-Alexandre Deslandres (1853-1948), Danjon became Director of Strasbourg Observatory (1930). The following year, he was also Professor of Astronomy at the Faculty of Sciences. In 1935, he became Dean of the Faculty and held those responsibilities until after the end of WWII in 1945. From 1940 to 1945, he ensured also the Vice-Presidency of the University Council that had moved to Clermont-Ferrand (Puy-de-Dôme). There Danjon had a few problems with the German authorities. He was emprisoned for some time and removed from office together with a few colleagues. The situation went back to normal in 1945, but Danjon did not return to Strasbourg as he was put in charge of the directorship at Paris Observatory on 1 May 1945.

That nomination was the result of the activity developed by Danjon at Strasbourg Observatory, *in situ* during almost ten years when he was Director there. While undertaking personal research, Danjon was granted in 1933 the construction of a new building of collective interest that he considered as a first institute where research in astrophysics would be carried out in France. Just before WWII, scientists such as André Lallemand, Paul Muller and others led there original investigations under the impetus of Danjon.

During his stay in Strasbourg, Danjon carried out investigations on observing methods and instruments. Those studies, as well as those by André Couder (1897-1979), who was for a short time in Strasbourg before working at the Laboratory of Optics in Paris, led to the book *Lunettes et télescopes* [Refracting and Reflecting Telescopes] authored by both. The volume appeared in 1935 for the first time, but has been republished several times, having become an indispensable reference in the field.

The French National Committee for Astronomy, set up shortly after the IAU creation and put under the authority of the Academy of Sciences, was then in full activity. It organized meetings, with invited foreign astronomers, where current problems and investigations were discussed. It was structured through national commissions, parallel to those of the IAU. Young promising astronomers attended them.

In 1945, when Danjon took over the directorship of Paris Observatory, he had the heavy task of putting back into operations not only that house, but also the various French observatories. They had certainly been somehow active during WWII, but only on basic terms and with a restricted personnel who had, as in the whole country, specific problems to be dealt with.

The research carried out by Danjon in Paris, the developments he undertook, as well as his ideas reaching maturity there, resulted largely from his earlier activities in Strasbourg and in Clermont-Ferrand where Strasbourg University had withdrawn.

Rougier (1886-1947) had joined Strasbourg Observatory at the same time as Danjon. Both collaborated and published together a paper on the green flash[4] as well as several studies on Saturn's Ring. It seems that Rougier called Danjon's attention on the double-image principle from which he and his collaborators took advantage. Rougier continued his career at Bordeaux Observatory from 1937 on.

[4]See the corresponding chapter by Heck (2005) in this volume.

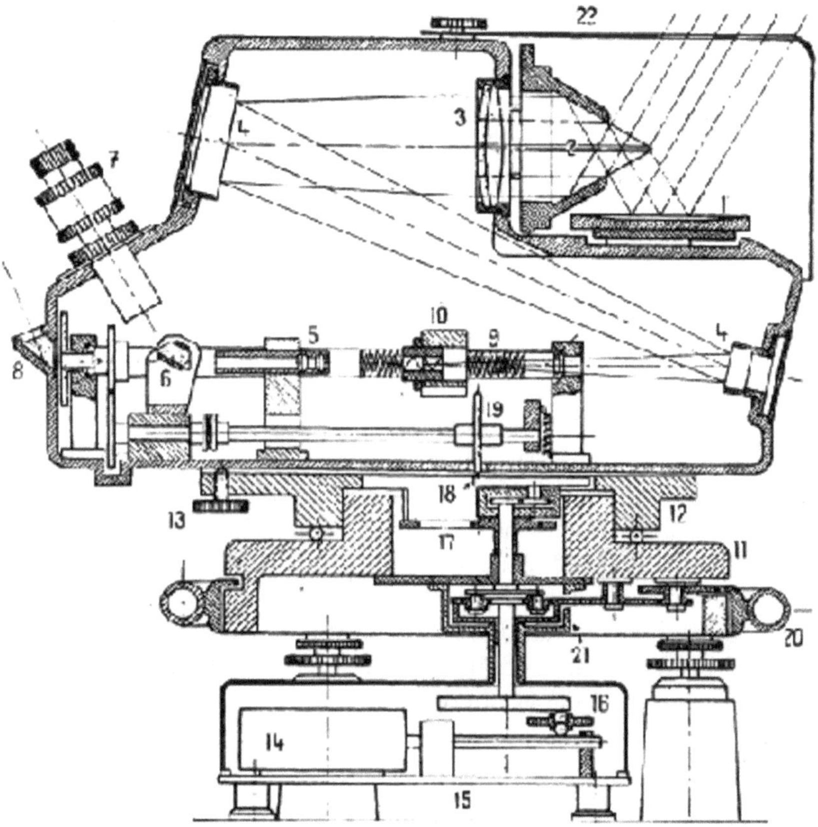

Figure 2. Danjon's astrolabe (Danjon 1958). (© Royal Astron. Soc.)

In Strasbourg, Danjon's interest increased for the precise determination of time, then based on the Earth rotation – two themes that would keep his thoughts busy during all his career. But they were not his only concern. He also got interested in double stars and in the celestial coordinates of various objects. Physicist by training, he attempted to solve many problems resulting from astrometry, a discipline without which it would be impossible to locate stars, nebulae and other objects in the universe.

Having already installed an astrophysical service in Strasbourg, Danjon, as explained above, played a decisive rôle in establishing the Observatoire de Haute-Provence (OHP), but also in setting up the European Southern Observatory (ESO).

Danjon had advocated constructing a 260cm telescope already in 1923 and later a 350cm one. One should not forget either that Danjon had introduced in the early 1950s a curriculum in astrophysics at the Sorbonne (Paris) for which he had enlisted a young *chargé de cours*, Evry Schatzman (b. 1920). Danjon had also been instrumental for renewing celestial mechanics in France, for developing computing in the observatories, for installing in 1956 – on land acquired in Nançay (Cher), following the recommendation of ENS' Y. Rocard – what would later become the Radioastronomy Station of Paris Observatory.

Another Strasbourg astronomer, Paul Muller, who had presented a PhD thesis on a novel instrument to measure double stars (see hereafter), was called by Danjon to Meudon Observatory in 1956. The following year, after tests carried out by several astronomers then at the Pic-du-Midi Observatory for observing the freshly launched first artificial satellite of our planet, Danjon created a service in charge of such a monitoring.

In Strasbourg like in Paris, Danjon gave great importance to teaching at various levels. Considering that one should not wait until after the *baccalauréat* [general certificate of education] to start learning astronomy, he had – as early as 1948 – conceived a remarkable course of cosmography for the level of elementary mathematics. This was announcing his textbook on *Astronomie Générale* [General Astronomy] published in 1952-1953. Danjon, who had been an educated amateur astronomer, did not spare his time either for helping amateur astronomers, creating in 1931 with others the Alsace Group of the *Société Astronomique de France (SAF).*

In the course of the quarter of century that he spent at Strasbourg Observatory, Danjon received a number of distinctions. He was awarded prizes from the Academy of Sciences several times between 1925 and 1938. Seriously wounded in Champagne, he received the *Croix de Guerre* [war cross] with palms as well as the *Légion d'Honneur* in 1915. Other honors would be subsequently bestowed on him.

He left his name to an astrometric instrument, the Danjon astrolabe, built in 45 copies installed in various countries. They were used for more than 30 years for the determination of time and latitude at those stations for the international services. Some of these are still in use today for other astrometric tests. Few professional astronomical instruments have been built to such a large number. He retired in 1963 and died on 21 April 1967.

Be it about Strasbourg or Paris or even other observatories from French provinces and/or abroad, Danjon could be characterized by a quote from one of his pupils (and there were many of them):

> "Ce fut l'un de ses plus grands mérites de comprendre immédiatement l'intérêt de toute nouveauté et d'en favoriser le développement, même si elle ne lui inspirait qu'un faible intérêt personnel." [5]

[5] "One of his biggest talents was to perceive immediately the advantage of any novelty

5. Lallemand: A Physicist serving Astronomy

André Lallemand was born on 29 September 1904 in Cirey-lès-Pontailler (Côte d'Or) of which he kept the local accent all his life. His university studies were conducted at Strasbourg University. In 1927, he graduated for teaching physical sciences. Two years earlier, he had been hired as assistant at the Observatory, then directed by Esclangon. He became then acquainted with Rougier and Danjon.

During several years, Lallemand could work also with Louis Néel (1904-2000), the future Nobel Prize winner. They had the same age and got both interested in the new methods for physics then in full development.

Lallemand started teaching at the *Lycée* of Haguenau in 1927-1928, before being nominated *Aide-Astronome* at Strasbourg Observatory, a position he held until 1938. In 1931, he was put in charge of lecturing on instruments until his departure for Paris Observatory.

Around 1933, Lallemand was frequently visiting Louis Hackspill's laboratory at the chemical institute where also André Couder worked, a renowned optician. This probably led Lallemand to set up an aluminizing plant for telescope mirrors, something he would also do later at the Paris Institute of Astrophysics and at Haute-Provence Observatory. Lallemand also learned at Hackspill's laboratory how to generate a high vacuum.

In 1928 at Strasbourg Observatory, Lallemand undertook research in photographic photometry in the near infrared. The following year he was observing a total solar eclipse during an expedition led by Danjon. Danjon had got acquainted to General Ferrié who coordinated the venture at the national level under the authority of the *Bureau des Longitudes (BdL)*. Rougier, another member of the expedition, intended to carry out photometric observations of the corona with a photometer built in Strasbourg, but these were hampered because of poor meteorological conditions at Poulo Condore (Indochina), the selected site.

On the other hand, Lallemand succeeded to conduct the first attempts of infrared photometry of the corona. He had indeed, the previous year, initiated research of photographic photometry in that wavelength range, a field he felt would expand. The photographs then obtained in the red and the infrared by Lallemand were certainly the first ones of the kind. He investigated the brightness of the stellar corona in several areas and at wavelengths of 672 and 830nm.

At Strasbourg Observatory, Lallemand took also his share in what is called today service tasks or activities of collective interest. He carried out time determinations and, in parallel, devised electronic relays enabling

and to favor its development, even if it triggered in himself only a limited personal interest."

to record the time of observations on printing chronographs. He studied equally the way fundamental clocks of the house were working. He improved their stability while pursuing, coached by Danjon, his photometric studies, both visual and photoelectric. Fruitful exchanges took place with Rougier who was working also on the photoelectric effect.

Rougier's successes prompted Danjon to orient Lallemand towards using this effect to devise a novel recorder of luminous energy, hoping it would be more powerful than those then in existence. Lallemand told one day one of his collaborators that he got the idea of the "electronic telescope" in a train between Strasbourg and Paris in 1934: in practice, it came to make operational what would be called later *caméra électronique* [electronic camera] or *caméra Lallemand* [Lallemand camera].

Such a device, based on magnetic focussing of photoelectrons, was first materialized at Strasbourg Observatory. The main interest resided in the fact that such a camera, put directly on a telescope, allowed a better resolution than photographic plates, with shorter exposure times and a linear response. In 1935, Lallemand was able to announce the first results in a note presented to the French Academy of Sciences by Esclangon on the application of electronic optics to photography. Two years later, the first experimental results were totally conclusive.

Lallemand had been encouraged by Jean Perrin (1870-1942), by Charles Fabry (1867-1945), by Louis de Broglie (1892-1987), by Armand de Gramont (1879-1962). He was promoted to *Astronome-Adjoint* at Strasbourg Observatory in 1938. In 1939, Lallemand constructed the first operational device with an electronic optics. But the war broke out at the end of that year. Lallemand and Néel were assigned to research services for the national defence. They worked to efficiently equip ships in charge of neutralizing magnetic mines. Lallemand tackled also the challenge of detecting low-intensity radiation.

Lallemand maintained links with navy people, such that, at the end of WWII, some navy services were assigned to Paris Observatory. The laboratory that Lallemand had created in Strasbourg had been totally destroyed and he had chosen to go to Paris Observatory already in 1942 where he was officially assigned in 1943.

Lallemand then devoted himself to studying photomultipliers for astronomy, having obtained in Strasbourg experimental electronic images requiring exposure times ten times shorter than by classical photography with the then so-called extra-rapid plates. Upon his arrival at Paris Observatory, Danjon set up for Lallemand a laboratory of astronomical physics centered on electronography.

In a few years, and beyond the collaborators joining him, the laboratory welcomed also a number of astronomers and physicists visiting the place for

Figure 3. Lallemand's electronic camera in a development by M. Duchesne at the Newton focus of the 193cm telescope of Haute-Provence Observatory in 1974. (© Haute-Provence Obs.)

documenting themselves or for being trained to the technologies developed. In 1952, Lallemand was nominated scientific councilor to the navy with the rank of *Ingénieur Général* [General Engineer]. The following year, he was promoted to *Astronome Titulaire* at Paris Observatory.

Soon cameras were developed, in collaboration with others, with a sensitivity a hundred times better than the most efficient photographic plates. This led Lallemand to write that: "Les idées directrices suivies à Strasbourg et qui m'avaient conduit à des résultats ont été reprises[6]." He clarified: "Il fallait rendre la méthode plus efficace et plus facile à utiliser [7] ...". In order to eliminate some secondary effects, he resumed collaboration with Néel, in the line of what had been done in Strasbourg well before.

Other cameras would soon see light with various collaborators, other cameras would be experimented in the course of the years, while Lallemand, having been nominated Professor at the *Collège de France* in 1951 to the chair of *Méthodes physiques pour l'astronomie* [Physical Methods for Astronomy], could say during his opening speech:

> "Si on me demandait pourquoi je me suis engagé à traiter ici des méthodes physiques de l'astronomie, je répondrais simplement: c'est parce que j'aime la physique et que je suis un astronome."[8]

That astronomer who loved physics retired from active life in 1974 and passed away on 24 March 1978.

6. Muller: An Adept of the Double-Image Technique

Paul Muller was born on 17 November 1916 in Lorquin (Moselle). Just after starting studies for a licence [bachelor's degree] in sciences at Strasbourg University, that amateur astronomer got a position as temporary assistant at the local observatory. This not only allowed him to continue his studies, but also gave him access to the astronomical professional circle and introduced him in a place where were working Rougier and Lallemand under the directorship of Danjon.

His degree secured in 1932, Muller got a (firm) position as Assistant in 1934. He went on training in professional observing while taking his share in the observatory activities. At his return in 1936 after one year spent in military duties and following a suggestion by Danjon, he tackled the study of the birefringence of images being at the root of the double-image technique.

With the help of Rougier and Danjon, Muller succeeded in designing a micrometer applying this principle to the measurement of double stars. He developed the related theory and, in 1938, the micrometer took its definitive shape. From the start, that novel instrument produced measurements three times more precise than the wire micrometers used so far.

[6] "The guiding ideas at Strasbourg that led me to results have been followed."
[7] "It was necessary to make the method more efficient and easier to be used."
[8] "If one would ask me why I committed myself to deal here with physical methods for astronomy, I would simply answer that I love physics and that I am an astronomer."

Muller had been nominated *Aide-Astronome* in 1938. In August 1939 though, he was mobilized, then made prisoner of war in June 1940. Rejecting the conditions for freedom offered to men from Alsace and Moselle, he then spent five years in offlag camps during which, every time he could, he perfected the theoretical treatise of his experimental studies. Muller was freed in April 1945.

Back to Strasbourg already in May 1945, he discovered that a big shell had destroyed the meridian room where, in 1939, Danjon had considered installing a new instrument using the double-image technique. While waiting for the return of astronomers and the nomination of a new Director (Lacroute, in 1946), Muller, almost alone, did his best to put the observatory back in operations, reinstalling the dismounted instruments with the help of Danjon, then already Director of Paris Observatory.

In 1945, Muller was promoted to *Astronome-Adjoint*. In 1947, he could finally present his PhD thesis in Paris. To this purpose, he carried out observations not only in Strasbourg, but also at the Pic-du-Midi and Haute-Provence Observatories. His thesis was published in 1948 under the title: *Sur un nouveau micromètre astronomique à double image, ses possibilités et quelques questions connexes* [On a New Astronomical Micrometer based on the Double-Image Technique, on its Capabilities and on a few Related Issues].

In different papers, Muller had indeed demonstrated the usage that could be made of that micrometer and that he made himself of it in the following years – first by measuring double stars, but also through photometric and colorimetric measurements (of the satellites of Jupiter and of their mutual phenomena, of Titan, of star occultations by the Moon, and so on).

Soon after, Muller tackled the orbit determination of double stars for which enough observations were available. Such stars are important as they are the only ones for which masses can be directly determined. Thus they can be used to calibrate the mass-luminosity relationships.

Having become a recognized world specialist for double stars, Muller was invited for four months at Lick Observatory (California) where he could use the local 91cm refractor. Thus, in 1953, he discovered there four new couples from the some 500 measurements he could conduct. In parallel, Muller undertook compiling a card index of double-star observations following those established by his predecessors. A catalogue of orbits was also put together.

Muller entered IAU Commission 26 (Double Stars) in 1946. He became its President in 1952 and, in 1954, he launched a *Circulaire d'information sur les étoiles doubles* [Information Circular on Double Stars] while a dozen observatories in the world adopted the double-image micrometer. Later on, Muller was again elected as IAU Commission 26 President, a fact seldom enough to be stressed here. Such a position allowed him to broaden the scope of cooperation with other *doublistes* [binarists] as they call often themselves. Those collaborations were then no more as restricted as at the beginning of his career when their extent was essentially European.

At the beginning of the 1950s, Muller was worried by the degree of neglect of the large French refractors for the measurement of double stars, the eye being able to distinguish components better than some automatic devices of the time.

Thus when Daniel Barbier (1907-1965) was detached to CNRS[9] and joined Paris Institute of Astrophysics, Danjon called up Muller to replace him at Paris Observatory. He was then transferred in 1956 to Paris, provisionally detached in a position of *Astronome Titulaire* that he would officially hold in 1960.

He then observed at the 38cm refractor of Paris Observatory, studying also the modernization of the Large Refractor (76cm) of Meudon Observatory. Danjon then transferred him to the service of the equatorial instruments responsible for those two refractors. The former one, called nowadays *Lunette Arago* [Arago Refractor], has a focal length of 900cm and is located in an outlet surrounded by trees. Its first objective lens became so damaged that it had to be replaced, which was done by another one of outstanding quality by the Henry brothers (1881).

Charles E. Worley (1935-1997), the US double-star specialist who requested testing the quality of the refurbished instrument, found it much better than expected as he could observe a few close binaries with a remarkable stability of the images. One can then understand why Muller got so much interested in that Paris refractor, entrusting it, with Danjon's blessing, to an amateur observer of double stars, Paul Baize (1901-1995), pedriatrician by profession, who conducted there numerous observations followed by a large number of orbit determinations and publications in professional journals.

Muller carried out his own observations mainly at the Meudon Large Refractor. In 1957, however, the first artificial satellite was launched and orbited the Earth. In Meudon, Muller managed to get the necessary equipment to observe those new objects. Supported by Danjon, he organized their systematic observing, first visual, then photographic. A team gradually took shape around Muller. When, fifteen years later, virtually the whole group joined the freshly created CERGA[10] North of Grasse (Alpes-Maritimes), Muller moved there too. He went then back to his first pet objects, the optical binaries, using this time the Large Refractor of Nice Observatory and the equatorial telescope located in the Charlois dome.

[9]CNRS = Centre National de la Recherche Scientifique.

[10]CERGA = Centre d'Études et de Recherches en Géodynamique et Astrométrie.

UNION ASTRONOMIQUE INTERNATIONALE

COMMISSION DES ETOILES DOUBLES

Février 1954

Circulaire d'Information N° 1

En Janvier 1953, le Prof.W.RABE (Munich) présentait un projet dont voici le texte original:

"Plan für ein Nachrichtenblatt der Doppelsternastronomie.

Zur Hebung des Interesses an der Doppelsternastronomie und zur leichteren und vollständigeren Unterrichtung über Ergebnisse und Erfordernisse, besonders in praktischer Hinsicht, scheint mir die Einrichtung eines Nachrichtenblattes etwa in der Form, wie es für kleine Planeten existiert, wünschenswert.

Ich denke mir, dass das Nachrichtenblatt nach Bedarf in unregelmässigen Intervallen erscheinen könnte, und folgende Angaben enthalten sollte:

1) vollständige Angabe aller neu erscheinender Bahnelemente und Ephemeriden;
2) Bibliographie aller einschlägigen Arbeiten;
3) Hinweise auf Sterne, die sich kritischen Bahnpunkten nähern;
4) evtl. persönliche Nachrichten und Anregungen der Kommission.

Das Nachrichtenblatt ist nicht zur Veröffentlichung von Beobachtungsreihen und Abhandlungen bestimmt, damit das Projekt nicht an den Finanzierungskosten scheitert."

Après consultation de plusieurs collègues, notamment lors d'un séjour aux Etats-Unis, et ayant dès mon retour pris avec le secrétaire général de l'U.A.I. les arrangements nécessaires, je me chargerai de composer et de diffuser de telles Circulaires, conformes presque à la lettre aux principes ci-dessus qui ont rencontré l'approbation générale. La Commission aura à décider de l'organisation définitive de ce service lors du Congrès de 1955.

Les Circulaires seront envoyées à tous les membres de la Commission 26, ainsi qu'aux observateurs actifs qui n'en feraient pas encore partie. Une limitation sévère permettra seule d'assurer l'acheminement par air, hors d'Europe, qui paraît inséparable de la rapidité d'information désirée. C'est pourquoi il ne sera envoyé qu'une circulaire aux Observatoires qui grouperaient plusieurs destinataires. J'ai arrêté de mon mieux la première liste d'envoi, mais je demande qu'il me soit suggéré des additions justifiées. Dans ce but, la création de ce service est annoncée également par la voie des Circulaires du Bureau Central des Télégrammes Astronomiques de l'U.A.I.

Pour les orbites communiquées avant publication, on donnera les éléments, une courte éphéméride, si possible la date de la dernière observation utilisée. La sélection des objets proposés en priorité aux observateurs sera plus délicate. Je crois cependant qu'il faudra s'adresser surtout aux instruments moyens (25 à 60 cm), assez largement

Figure 4. The first issue of the Information Circular on double stars launched in 1954 by Paul Muller from Strasbourg for Commission 26 of the International Astronomical Union. Notice the text in German reproduced without translation.

As to artificial satellites, Muller had convinced his colleagues from Besançon, Bordeaux and Strasbourg Observatories to somehow set up an observing network that was joined in 1960 by the overseas stations of the national meteorological service. In Strasbourg and Meudon, the observations started rather quickly with the help of the US Air Force. In the same field of artificial satellites, Danjon got for Muller a position at the French Committee for the International Association of Geodesy (IAG), as he did also at the creation of COSPAR[11] in 1958.

All those steps, from Danjon and Muller who both started their career at Strasbourg Observatory, as well as from other scientific personalities, led to the birth of the *Centre National d'Études Spatiales (CNES)* [French Space Agency] created in 1962.

7. Conclusion

Among astronomical observatories in France, Strasbourg Observatory offers several peculiarities linked to the context of its creation, of its location and of its evolution during the first half of century that ended in 1920.

The novel ideas that have been developed at Strasbourg Observatory in the 1930s and the multidisciplinary exchanges that took place there have marked the astronomical instrumentation of many observatories around the world for more than another half of century, with some uses that went into the 21st century.

Sources

Here are the main sources for the above:

- Capitaine, N. & Débarbat, S. (Eds.) 1990, Colloque André Danjon – Systèmes de Référence Spatio-Temporels, Obs. Paris, x + 310 pp. (ISBN 2-901057-20-9)
- Danjon, A. 1958, The Contribution of the Impersonal Astrolabe to Fundamental Astronomy, *Monthly Not. Royal Astron. Soc.* **118**, 411-431.
- Danjon, A. 1965, Courte Histoire de l'Observatoire de Haute-Provence, Obs. Paris, 32 pp.
- Duerbeck, H.W. 2005, The Observatory of the Kaiser-Wilhelm University: The People Behind the Documents, this volume.
- Esclangon, E. 1926, La Nouvelle Organisation de l'Observatoire, *Ann. Obs. Strasbourg* **I**, 1-44.
- Heck, A. 2005, Strasbourg Green Rays, this volume.

[11] COSPAR = Committee on Space Research.

- Weidman, C., Twoney, S. & Huffman, D.R. 1990, Variations in Earthshine Due to Changes in Reflection by the Earth's Cloud Cover, in *Colloque André Danjon – Systèmes de Référence Spatio-Temporels*, Eds. N. Capitaine & S. Débarbat, Obs. Paris, 237-238.
- Wolfschmidt, G. 2005, Strasbourg Observatory in German Times, this volume.
- Personal activity reports from A. Danjon, E. Esclangon, A. Lallemand and P. Muller, Paris Obs. library.
- Activity reports for Paris Observatory covering the period 1928-1940.
- Personal communications from Léopold Renard.
- Personal memories from the author.

Acknowledgments

The author is grateful to the Editor for bringing some additional pieces to the orginal manuscript and for translating this contribution into English.

WALTER F. WISLICENUS AND MODERN ASTRONOMICAL BIBLIOGRAPHY

H.W. DUERBECK
Vrije Universiteit Brussel
Pleinlaan 2
B-1050 Brussel, Belgium
hduerbec@vub.ac.be

Abstract. A short overview of the bibliography of astronomy from the time of Weidler (1755) to the end of the 19th century is given. Walter Wislicenus studied astronomy in Leipzig and Strasbourg, took part in the German Transit of Venus expedition to Bahia Blanca in 1882, finished his PhD, became an assistant at Strasbourg Observatory, and afterwards a private lecturer and professor of astronomy. Because of personal reasons, his connection with the observatory became very loose, and he focussed on writing books on various astronomical topics, from chronology over geodesy to astrophysics. His last and major enterprise, the *Astronomischer Jahresbericht (AJB)*, whose first six volumes he compiled and edited, survived in different forms for more than 100 years.

1. Astronomical Bibliography Throughout the Ages

Only in the late 17th century, the first systematic compilations of astronomical literature began to appear. The first important historiograph of astronomy of modern times was also the first editor of an astronomical bibliography: Johann Friedrich Weidler (1692-1755), a professor of mathematics and legal studies at Wittenberg university.

His *Historia Astronomiae* was published in 1741, his *Bibliographia Astronomica* in 1755. Arranged according to the year of publication, the latter one contained about 1200 references to works written or published from 480 B.C. to 1755. This was followed by an *Astronomische Bibliographie*, which constitutes the third part of a more extensive *Einleitung zur Mathematischen Bücherkenntniß*, published 1769-1798 by Johann Ephraim Scheibel (1736-1809), a mathematics and physics professor in Wroclaw [Breslau].

A. Heck (ed.), The Multinational History of Strasbourg Astronomical Observatory, 153–165.

This compilation contains 1600 references of books published between 1400 and 1650.

The 19th century saw two major bibliographies; in 1803, Jérôme de Lalande (1732-1807) published his *Bibliographie astronomique*, another chronological listing of 5200 books, which is the last complete astronomical bibliography of all times and subjects that fits within the covers of a single volume. In addition, two specialized bibliographies appeared: a listing of writings edited by scholarly societies in 16 volumes, which appeared under the title: *Repertorium Commentationum a Societatibus Litterariis Editarum, Secundum Disciplinarum Ordinem*, Gottingae 1801-1821.

It was compiled by Jeremias David Reuß (1750-1837), a professor of philosophy and history of literature, and sub-librarian of Göttingen University. This bibliography is restricted to journal literature published by scholarly societies, but because most scientific journals in the 17th and 18th centuries were issued by scholarly societies, this is a good index to scattered articles that had appeared until that period. The fifth volume, *Astronomia*, issued in 1804, comprises astronomical papers written till about 1800.

Another source is the *Handbuch der Deutschen Literatur* by Johann Samuel Ersch (1766-1828), whose second part of the 3rd volume (*Literatur der Mathematik, Natur- und Gewerbskunde mit Inbegriff der Kriegskunst und anderer Künste, außer der Schönen*) appeared in 1828. The reference period is 1750 – 1825, astronomy covers columns 204-231 and 1099-1107.

The century came to an end with the largest of all printed astronomical bibliographies, which unfortunately remained uncompleted. Jean-Charles Houzeau (1820-1888), director of the Royal Observatory in Brussels, and Albert Lancaster (1849-1908), his librarian, started to compile a complete bibliography, which appeared in several installments between 1880 and 1889. This *Bibliographie Générale de l' Astronomie* was planned to comprise three volumes:

I. printed books and manuscripts,
II. scientific papers in academy publications and journals, and
III. astronomical observations.

Volume II appeared between 1880 and 1882, three parts of the first volume, covering historical books, biographies, textbooks, books on spherical and theoretical astronomy, appeared in 1886, 1887 and 1889. The bibliography comprises about 20000 titles. Houzeau died in 1888, and the project remained unfinished.

2. Walter F. Wislicenus

On the basis of his curriculum vitae, given in Wislicenus (1886), and several obituaries (Anonymous 1905, Becker 1905, Kobold 1906), we gather

Figure 1. Walter F. Wislicenus (portrait from Vol. 7 (1906) of the *Astronomischer Jahresbericht*).

the following information: Walter Friedrich Wislicenus (Fig. 1) was born in Halberstadt (Prussian Province of Saxonia) on 5 November 1859, the son of the preacher of the local free-religious community, Adolf Timotheus Wislicenus, and his wife Helene Wislicenus, née Menzzer. In Autumn 1861, the family moved to Berlin, where he attended the preparatory classes of a secondary school [Gymnasium] from Autumn 1866 to Summer 1868. When his family moved to Dresden, he entered an educational institute from 1868 to 1870; afterwards he attended secondary school classes in Dresden, and obtained his diploma [Zeugnis der Reife] on Easter 1879.

He studied mathematics and astronomy, in his first year at Leipzig University, where his teachers in astronomy, mathematics and physics were

Carl Christian Bruhns, Wilhelm Scheibner, Karl Friedrich Zöllner and Hugo Seeliger. At that time, Leipzig Observatory was in charge of the analysis of photographic observations of the Venus transit of 1874; but in the course of time, it had turned out that the heliometer observations, supervised by August Winnecke, the director of Strasbourg Observatory, yielded superior results. This may have been one reason that, starting Easter 1880, Wislicenus continued his studies at Strasbourg University. Here he first continued theoretical studies. In December 1881, he applied to participate in one of the German Venus transit expeditions of 1882 (Fig. 2), and was soon accepted[1].

In Spring 1882, he started to make heliometer observations, to prepare for the task. At the end of 1882, he participated as "third astronomer" in the third German expedition for the observation of the Venus transit, which was bound to Bahia Blanca in Argentina (Fig. 3). The team obtained some useful observations in spite of unfavorable weather (Duerbeck 2004). He returned to Strasbourg in February 1883, soon became secondary assistant ["Hülfsassistent"] at the Observatory, and assistant on 1 April 1884, a post that he kept until 1889.

In February 1885, he interrupted his observational activity to start work on his Ph.D. dissertation. This work, completed by the end of 1885, is dedicated to the memory of his father, who had died shortly before, and bears the title "A contribution to the determination of the time of rotation

[1]His application letter runs as follows:

Strasbourg, Alsace, 18 December 1881.

Very esteemed Sir,
since it is said that the German Empire will furnish several expeditions to observe the Venus transit in front of the solar disk on 1882 December 6, I dare to offer my services and my person to a highly esteemed commission, which is in charge of the expedition. In spite of the fact that my astronomical studies, to whom I apply myself since three years, have not yet been carried out to the end, I nevertheless hope that I could be of some use to the esteemed commission when possibly participating in one of the expeditions. In the presumably quite long time till the departure of the expeditions, I hope to be able to fill all gaps in my knowledge and skills which are necessary for such an expedition, and at the same time I herewith promise to carry out all necessary steps, as far as they depend on myself, as soon as possible. The most important lack which has to be removed from my side is my ignorance of heliometric measurements and observations, but I hope that I will soon surmount this difficulty, since I intend to start, as soon as weather permits, under your excellent guidance and instruction at the Imperial University Observatory of Strasbourg, with the heliometer observations. Finally, by promising to an esteemed commission to obey most readily and punctually to each of its instructions, I beseech you, highly esteemed Sir, to kindly put before an esteemed commission this offer of mine to its complaisant information.
By assuring you of my most humble respect, I remain yours faithfully
Walter Wislicenus,
student of astronomy at the Emperor Wilhelm University Strasbourg.
Strasbourg, Alsace, Feggasse 16, II.

Strassburg im Elsass 1881 December 18.

Hochgeehrtester Herr!

Walter Wislicenus

Figure 2. Letter of W.F. Wislicenus applying for participation in the transit of Venus expeditions of 1882. The letter was written and "received" 1881 December 18 (stamp and abbreviated signature of Winnecke) and forwarded to the head of the German Transit of Venus Commission, Auwers (stamp and date 1881 Dec. 20). Some marginal notes by Auwers obviously deal with the answering of the letter and renumeration (Archive of the Berlin-Brandenburg Academy of Sciences, Berlin).

Figure 3. Section of a photograph of the German observing station in Bahia Blanca, Argentina, to observe the Venus transit of 1882. Standing near the telescope are, from left to right, Wislicenus, mechanic Meyer and astronomer Peter, while the head of Hartwig, the expedition leader, is visible in the dome slit (Archive of the Berlin-Brandenburg Academy of Sciences, Berlin).

of the planet Mars" (Wislicenus 1886). A 26-page collection of historical observations of surface markings, which range from Fontana's 1636 to 1638 observations, to publications of 1884, is used to derive the time of rotation, which he determined to be 24 hours, 37 minutes 22.655 ± 0.00861 (m.e.) seconds. A similar result was obtained almost simultaneously by Hendricus Gerardus van de Sande Bakhuyzen (1885): 24 hours, 37 minutes 22.66 ± 0.0132 seconds (the present day value is 22.663 seconds).

In 1884, when Ernst Hartwig left Strasbourg to become astronomer in Dorpat, Wislicenus took over the assistant position. In 1886, when Wilhelm Schur moved to Göttingen, Wislicenus hoped to be appointed to the vacant observer's position, but the Berlin ministry, obviously on the advice of Arthur Auwers, called Hermann Kobold, an astronomer working for the German Commission for the observation of the Venus transits, to the post.

The personal contact with van de Sande Bakhuyzen led Wislicenus also to the study of the personal equation, and he wrote his habilitation thesis about this topic in late 1887 (Wislicenus 1888), while again a parallel investigation was published by van de Sande Bakhuyzen (1889). With this thesis,

Wislicenus obtained the *venia legendi* (teaching license for the university). Although he stepped down from the assistant position at the observatory on 1 October 1889, he remained a private lecturer [Privatdozent, without formal payment by the university]. In 1894, he became extraordinary professor at Strasbourg University.

The official records do not mention why he quit his position, but it seems that continuing intrigues between him – who had some good relations with faculty members and the administration of the university – and observatory staff (first his competitor Kobold, finally the director, Ernst Becker) lead to his dismissal on 1 October 1889. Some comments are given in Kobold's personal notes (Kobold 2004); it is interesting to note that the two obituaries, written by Becker (1905) and Kobold (1906), hardly give a hint to these quarrels. Wislicenus continued to observe, but was not any more associated with the ongoing work of the institution; he had to ask the director for observing time for his own research programs. Among his observational work, there are some sparse observations of Mars, and a series of selenophotometric observations, carried out between 1893 and 1896, and edited posthumously by Carl Wirtz (1915) (Fig. 4)[2].

Wilhelm Foerster (1911) writes in his memoirs: *Professor Wislicenus belongs to the most zealous and most active collaborators of the "ethic movement"* [Ethische Bewegung]. So how can this be reconciled with the reproach of being an intriguer? Let us quote a few lines from Kobold's (1906) obituary: *His external appearance was, in spite of his politeness, in general a reserved one, and he appeared to be arrogant and suspicious to persons who did not know him well. However, his inner being, to which only a few could penetrate, was quite pure and modest. It was as if two natures would strug-*

[2]The letter written by C. Wirtz to Mrs. Wislicenus reads:

Imperial University Observatory — Strasbourg in Alsace, 22 October 1915.

Very esteemed Madam,
Please permit me to present you the finished reduction of the valuable observations of your husband. I received the material 10 years ago, in order to subject it to a precise calculation. It may appear that I have neglected this esteemed duty of thankfulness and reverence because of pressing other studies. Meanwhile, my friend H. Rosenberg in Tübingen had been stimulated by Wislicenus' moon photometry to studies whose results I would have liked to compare with those presented here. In the meantime the war broke out, Rosenberg is at the front, and in order not to postpone the publication at random, I edited the study in the form that you have before your eyes.

I have to fulfill the pleasant duty to recall the merits of friends and colleagues of the revered deceased for the appearance of the moon photometry. Mr. Kobold gladly accepted the paper for the Astronomische Nachrichten, in spite of its large size, and Mr. Bauschinger paid the not unsubstantial cost for the clichés and tables from the budget of the observatory.

With cordial greetings,
Your most obedient
Carl Wirtz

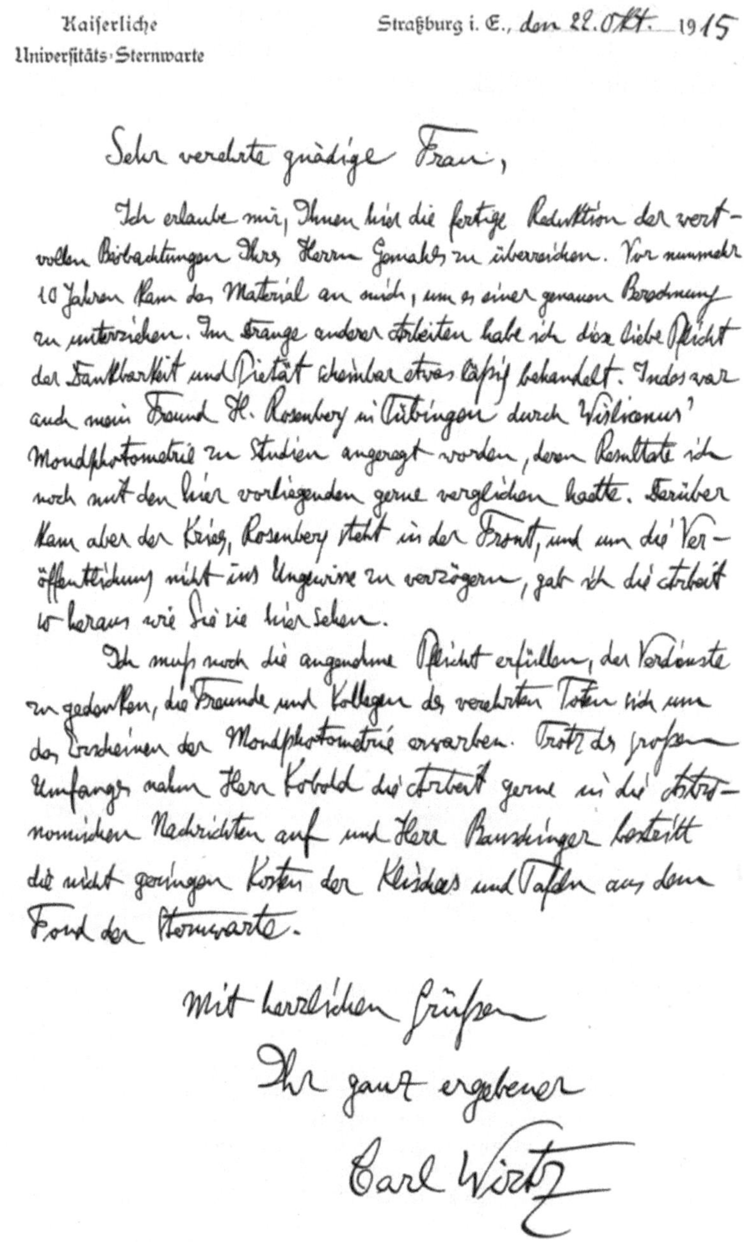

Kaiserliche
Universitäts-Sternwarte

Straßburg i. E., den 22. Okt. 1915

Sehr verehrte gnädige Frau,

Ich erlaube mir, Ihnen hier die fertige Reduktion der wertvollen Beobachtungen Ihres Herrn Gemahls zu überreichen. Vor nunmehr 10 Jahren kam das Material an mich, um es einer genauen Berechnung zu unterziehen. Im Drange anderer Arbeiten habe ich diese liebe Pflicht der Dankbarkeit und Pietät scheinbar etwas lässig behandelt. Indes war auch mein Freund H. Rosenberg in Tübingen durch Wislicenus' Mondphotometrie zu Studien angeregt worden, deren Resultate ich noch mit den hier vorliegenden gerne verglichen hätte. Darüber kam aber der Krieg, Rosenberg steht in der Front, und um die Veröffentlichung nicht ins Ungewisse zu verzögern, gab ich die Arbeit so heraus wie Sie sie hier sehen.

Ich muss noch die angenehme Pflicht erfüllen, der Verdienste zu gedenken, die Freunde und Kollegen des verehrten Toten sich um das Erscheinen der Mondphotometrie erwarben. Trotz des grossen Umfanges nahm Herr Kobold die Arbeit gerne in die Astronomischen Nachrichten auf und Herr Bauschinger bestritt die nicht geringen Kosten der Klischees und Tafeln aus dem Fond der Sternwarte.

Mit herzlichen Grüssen
Ihr ganz ergebener
Carl Wirtz

Figure 4. Letter of C. Wirtz written to Mrs. Wislicenus concerning the edition of a posthumous article (private collection).

gle inside him, one that would restlessly force him forward, made him go round all obstacles or defeat them, and another one that was introverted, that would gladly enjoy the measure of happiness that was acquired by diligence and constancy.

After being a "free" person, his scientific activity covered many topics, and his literary activity was large: he wrote books on position-finding (Wislicenus 1891), astronomical chronology (Wislicenus 1895), general astronomy (Wislicenus 1898a, 1903), astrophysics (Wislicenus 1899), and the calendar (Wislicenus 1905). But his most recognized astronomical achievement is the editorship of the *Astronomischer Jahresbericht (AJB)*, which will be discussed in Sect. 3. Six volumes, covering the years 1899 to 1904, were edited by him.

A typhoid infection led to a rapid death, he died on 3 October 1905, being almost 46 years of age. His idea and his realization of a printed annual bibliography was carried forth for more than a century, first by the German-language *Astronomischer Jahresbericht* (till 1968), then by *Astronomy and Astrophysics Abstracts*, written in English (1969-2000).

3. Wislicenus' annual astronomical bibliography

A "literary enterprise", which "should give a scientific annual overview in the complete field of astronomy", appeared for the first time in 1900: The *Astronomischer Jahresbericht, herausgegeben durch Walter F. Wislicenus, enthaltend die Litteratur des Jahres 1899* (Fig. 5). A handful of collaborators, mostly working in foreign countries, assisted him in cataloguing the publications which appeared all over the world, and in most cases wrote a short summary (or abstract).

The plan and the principles of this project were announced during the 1898 meeting of the Astronomische Gesellschaft in Budapest (Wislicenus 1898b), and information was also published in the Astronomical and the Astrophysical Journals (Wislicenus 1899).

Wislicenus was able to edit the first six volumes (1899-1904). After his death, Adolf Berberich (1861-1920) served as an editor until Vol. 11 (1909), when failing health made a continuation impossible, and Fritz Cohn (1866-1922) took over. Shortly before, Cohn had become director of the Astronomisches Rechen-Institut Berlin, and the previously privately organized project was incorporated into the activities of the institute. Cohn had prepared Vol. 12 (1910), and later prepared the years 1916-1921, Paul V. Neugebauer was in charge of the years 1911-1915 and 1922-1933.

His task was taken over by Karl Heinemann (1934-1958), and later by Werner Lohmann (1959-1968). A description of the project is found in Heinemann (1953). Lohmann (1962) gives an update, mentioning also

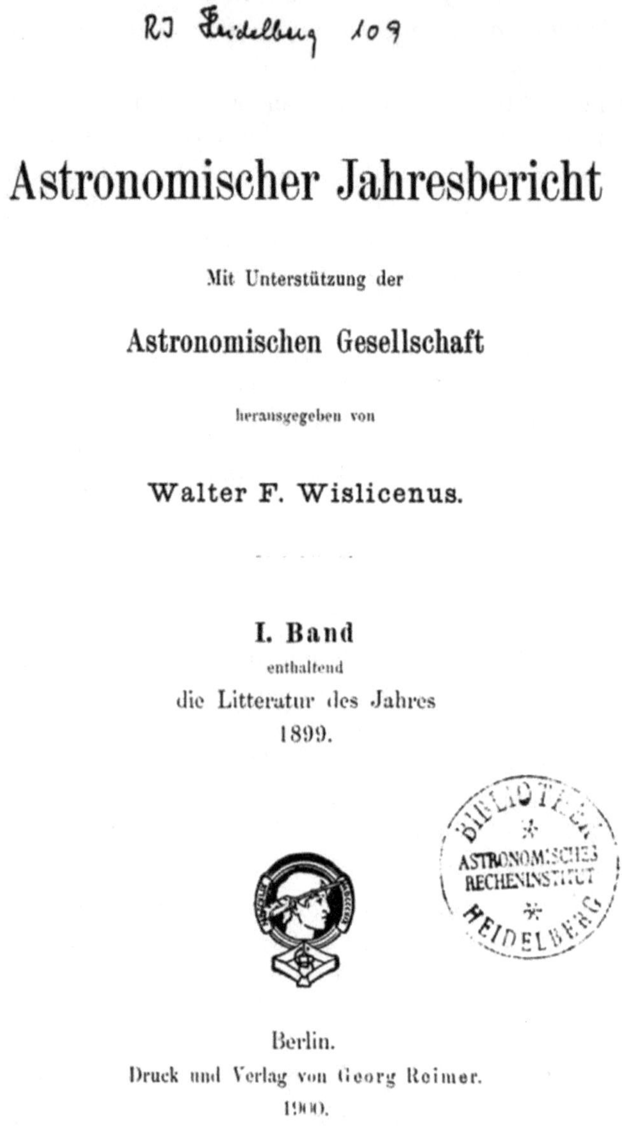

Astronomischer Jahresbericht

Mit Unterstützung der

Astronomischen Gesellschaft

herausgegeben von

Walter F. Wislicenus.

I. Band

enthaltend

die Litteratur des Jahres

1899.

Berlin.

Druck und Verlag von Georg Reimer.

1900.

Figure 5. Title page of the first volume of the *Astronomischer Jahresbericht.*

the (then) concurring projects. In Wislicenus' days, a concurrent project, the *International Catalogue of Scientific Literature, Sect. E (Astronomy)*, published by the Royal Society, was stopped in the World War I. In France, a monthly bibliography was issued, the *Bibliographie mensuelle de l'astronomie* (1933-1943) and its sequels, the *Bulletin analytique* (1940-1953) the *Bulletin signalétique-Astronomie et astrophysique* (1953-1983), and *Pascal explore* (1984-1989). In the Soviet Union, the *Referativny Zhurnal*, subsection Astronomy, has been issued from 1953 onward.

The *Astronomischer Jahresbericht* underwent a major reformation in 1969, it was replaced by the half-yearly *Astronomy and Astrophysics Abstracts* (AAA); the principally German-written publication turned into an English one (Schmadel 1979). In spite of modern publication methods, the growing size and content of the books slowed down the publication process, and made the consultation of many author and keyword registers somewhat tedious, in spite of the fact that index volumes, covering about 10 year intervals, were also prepared. The rise of at first less complete, but permanently updated and extended online databases, mainly the Astrophysics Data System (ADS)[3], made the AAA more and more obsolete. Finally, financial constraints of the State of Baden-Württemberg, which at present finances the Astronomisches Rechen-Institut, led to discontinuing AAA publication in 2000.

In any case, Wislicenus' project lasted for 100 years, and materialized into a set of 130 paper volumes (plus a few index volumes) that will hopefully survive the next millennium. But it is clear that astronomical bibliography in the 21st century has to follow other tracks.

It remains to be mentioned that the years 1881 to 1898 are, according to the information given above, poorly documented. Commission 5 of the International Astronomical Union (Astronomical Documentation) was actively engaged in bridging this gap. A card catalogue was prepared at the Royal Observatory Brussels, but financial means to typeset and print it were lacking, a microfilm of the catalogue (Sykes 1970) is only available in a few places. Attempts to make it – as well as the contents of the AJB and AAA – computer readable and generally accessible are still a desideratum of astronomical bibliography.

Sources and Acknowledgments

The general outline of astronomical bibliography, as outlined here, is based on the articles of Heinemann (1953), Lohmann (1962), Duerbeck (1977) and Schmadel (1979). My thanks go to the Archive of the Berlin-Brandenburg Academy of Sciences (Berlin) for providing Figs. 2 and 3. I sincerely thank

[3]`http://adsabs.harvard.edu/`

Brenda Corbin, librarian at the USNO, Washington, D.C., for supplying bibliographical information.

References

1. Anonymous 1905, Gedächtnisreden, gehalten bei der Trauerfeier für Prof. Walter F. Wislicenus, J.H.E. Heitz, Straßburg.
2. Becker, E. 1905, Todes-Anzeige [Walter F. Wislicenus], *Astron. Nachr.* **169**, 301.
3. Duerbeck, H.W. 1977, Astronomische Bibliographie, *Sterne und Weltraum* **16**, 241.
4. Duerbeck, H.W. 2004, The German Transit of Venus Expeditions of 1874 and 1882, *J. Astron. History and Heritage* **7**, 8.
5. Foerster, W. 1911, Lebenserinnerungen und Lebenshoffnungen (1832 bis 1910), Georg Reimer, Berlin.
6. Heinemann, K. 1953, Astronomische Bibliographie, *Die Sterne* **29**, 33.
7. Kobold, H. 1906, Walter Wislicenus [Nekrolog], *Vierteljahrsschrift Astron. Ges.* **41**, 13.
8. Kobold, H. 2004, Blätter der Erinnerung/Leaves of Memory, *J. Astron. Data* **10**, 5B.
9. Lohmann, W. 1962, 60 Jahre "Astronomischer Jahresbericht", *Sterne und Weltraum* **1**, 173.
10. Möbius, A.F. 1903, Astronomie – Größe, Bewegung und Entfernung der Himmelskörper. 10. verbesserte Auflage, bearbeitet von Dr. Walter F. Wislicenus. G.J. Göschen'sche Verlagshandlung, Leipzig.
11. Schmadel, L.D. 1979, 10 Jahre "Astronomy and Astrophysics Abstracts", *Sterne und Weltraum* **18**, 294.
12. Sykes, J.B. 1970, Guide to the Bibliography of Astronomy 1881-1898 on Microfilm, University Microfilms Ltd., High Wycombe.
13. van de Sande Bakhuyzen, H.G. 1885, Untersuchungen über die Rotationszeit des Planeten Mars und über Aenderungen seiner Flecke, *Ann. Sterr. Leiden* **7**, 1 (esp. p. 67).
14. van de Sande Bakhuyzen, H.G. 1888, Beschreibung eines Apparats zur Bestimmung des absoluten persönlichen Fehlers..., *Ann. Sterr. Leiden* **7**, 75.
15. Wirtz, C. 1915, W.F. Wislicenus' selenophotometrische Beobachtungen, *Astron. Nachr.* **201**, 289.
16. Wislicenus, W.F. 1886, Beitrag zur Bestimmung der Rotationszeit des Planeten Mars (Inaugural-Dissertation Strassburg), Braunsche Hofdruckerei, Karlsruhe.
17. Wislicenus, W.F. 1888, Untersuchungen über den absoluten persönlichen Fehler bei Durchgangsbeobachtungen, W. Engelmann, Leipzig.
18. Wislicenus, W.F. 1891, Handbuch der geographischen Ortsbestimmung auf Reisen, W. Engelmann, Leipzig.
19. Wislicenus, W.F. 1895, Astronomische Chronologie. Ein Hülfsbuch für Historiker, Archäologen und Astronomen, B.G. Teubner, Leipzig.
20. Wislicenus, W.F. 1898a, Möbius, A.F.: Astronomie – Größe, Bewegung und Entfernung der Himmelskörper. 9. verbesserte Auflage, bearbeitet von Dr. Walter F. Wislicenus, G.J. Göschen'sche Verlagshandlung, Leipzig (reprinted 1900).
21. Wislicenus, W.F. 1898b, Ueber die Herausgabe eines Astronomischen Jahresberichts, *Vierteljahrsschrift Astron. Ges.* **33**, 257.
22. Wislicenus, W.F. 1899, Communication Concerning the Publication of an Annual Astronomical Report, *Astron. J.* **29**, 16
   – An identical text is available in: An Annual Report on the Progress of Astronomy, *Astrophys. J.* **9**, 118.
23. Wislicenus, W.F. 1899, Astrophysik – die Beschaffenheit der Himmelskörper, G.J. Göschensche Verlagshandlung, Leipzig (First edition 1899, second, improved edition 1903, editions revised by H. Ludendorff appeared in 1909 and 1920).

24. Wislicenus, W.F. 1900, Bericht über den Astronomischen Jahresbericht, *Vierteljahrsschrift Astron. Ges.* **35**, 307.
25. Wislicenus, W.F. 1902, Bericht über den Astronomischen Jahresbericht, *Vierteljahrsschrift Astron. Ges.* **37**, 183.
26. Wislicenus, W.F. 1903, Möbius, A.F.: Astronomie – Größe, Bewegung und Entfernung der Himmelskörper. 10. verbesserte Auflage, bearbeitet von Dr. Walter F. Wislicenus, G.J. Göschen'sche Verlagshandlung, Leipzig (reprinted 1905).
27. Wislicenus, W.F. 1904, Bericht über den Astronomischen Jahresbericht, *Vierteljahrsschrift Astron. Ges.* **39**, 158.
28. Wislicenus, W.F. 1905, Der Kalender in gemeinverständlicher Darstellung, B.G. Teubner, Leipzig (Aus Natur und Geisteswelt).

THE NEBULAR RESEARCH OF CARL WIRTZ

H.W. DUERBECK
Vrije Universiteit Brussel
Pleinlaan 2
B-1050 Brussel, Belgium
hduerbec@vub.ac.be

AND

W.C. SEITTER
Universität Münster
D-48149 Münster, Germany

Abstract. We briefly review the nebular research in the 19th and early 20th century, and the rôle played by the newly founded Strasbourg Observatory in this field. The life of Carl Wirtz (1874-1939) is outlined. His studies in Bonn, his work in Vienna and Hamburg, his astronomical activity in Strasbourg from 1903-1916, at the war headquarters in Berlin in the following two years, and from 1919 up to his forced retirement in 1937 at Kiel University are described, both in general terms and specifically in terms of extragalactic research. His achievements were rarely recognized by his contemporaries, both because of his somewhat unusual way of presentation, and because of his "inner emigration" in the last years of his life.

1. Introduction

Nebulae have fascinated astronomers for centuries. In the 18th century, Messier and his contemporaries compiled catalogues of nebulae as a convenient way to avoid confusion with newly discovered comets. William Herschel and his family catalogued nebulae and star clusters as inhabitants of the sidereal system. Some keen thinkers like Thomas Wright of Durham or Immanuel Kant took them for systems similar to our Milky Way, an opinion that influential writers like Agnes Clerke or astronomers like Harlow Shapley tried to refute till the 1920s. How to deal with them? For a long time, telescopes were too small to resolve them into single stars. In 1845, Lord

A. Heck (ed.), The Multinational History of Strasbourg Astronomical Observatory, 167–187.

Rosse's *Leviathan of Parsonstown*, a 1.83-m reflector, provided first hints of spiral structure in some nebulae. This was followed by the discovery of many more "spiral nebulae" on photographic plates. In the closing years of the 19th century, early pioneers of this photographic study of nebular morphology were Isaac Roberts, an amateur astronomer in England, and James Keeler, director of the Lick Observatory in California, a task soon taken up by Max Wolf of the [then] Astrophysical Institute Heidelberg-Königstuhl, by George W. Ritchey at Yerkes Observatory, and by others.

Early astronomical spectroscopes were too inefficient to reveal the spectrum of these nebulae; painstaking, but inconclusive photographic and visual observations of the spectrum of the Great Andromeda Nebula are, *e.g.*, described in the *Atlas of Representative Stellar Spectra* by Huggins & Huggins (1899). The first convincing photographic spectrogram of the Andromeda nebula was taken in early 1899 by Julius Scheiner (1899) in Potsdam, using a specially constructed reflector and spectrograph; it revealed a solar-type spectrum. Only ten years later, Edward Fath (1909) at Lick and Max Wolf (1909) in Heidelberg managed to take spectra of additional spiral nebulae. Even then, the cosmic role of nebulae was not yet clarified: did they belong to the Milky Way, and did they share the motions of the stars? Or were they objects beyond the Milky Way?

At the turn of the century, when Wirtz started his work at Strasbourg Observatory, no Hertzsprung-Russell diagram was known, no distance indicator like a pulsating star or nova was readily available. There appeared to exist only one way to find the place of nebulae in the cosmos: to study their proper motion, to find their secular parallax, and to determine their apex in the sky, tools that had been used by astronomers in the previous decades to determine properties of various groups of stars that were, of course, members of the vast system of the Milky Way.

When August Winnecke was appointed director of the newly founded observatory of Strasbourg University, he stated that the principal research program would be "the study of the nebulae" (Winnecke 1875). His instrumentation was streamlined for this type of research: a large refractor with small focal ratio, to clearly examine and measure nebulae (*i.e.* their positions relative to comparison stars), and a meridian circle, to determine accurate positions of these comparison stars. Winnecke was a classical astronomer with classical research tools and methods: micrometrically measured positions, parallaxes, and proper motions – and he simply did not recognize, as most astronomers of his generation, the potential of photography and spectroscopy, methods mainly introduced into astronomy by such "amateurs" like Isaac Roberts and William Huggins in England, Henry Draper in the United States, and Nikolaus Thege von Konkoly in Hungary. Such methods were increasingly applied also in specialized observatories like the Astrophysikalisches Observatorium Potsdam (which was founded at about the same time as Strasbourg Observatory), and subsequently pursued at observatories in the United States like Lick, Mount Wilson, Lowell, and – with a different direction of research – at the observing stations of the Smithsonian Astrophysical Observatory.

The harvest of Winnecke's project, his preparatory observations, and the nebular observations made by him and by the observers Wilhelm Schur, Hermann Kobold and Carl Wirtz with the large refractor were published – about 35 years after its incipience! – in three parts: Sections 1 & 2 appeared in Vol. 3 of the Strasbourg *Annalen* (Becker 1909, Wirtz 1909) while Section 3, nebular observations by Wirtz, was published in Vol. 4, Part 1 (Wirtz 1911a).

Section 3 was followed by a general catalogue of nebulae observed in Strasbourg (Wirtz 1911b) and by a comparison of the Strasbourg general catalogue with other lists of nebulae (Wirtz 1911c). In the following Part 2 of Vol. 4, Wirtz (1912) included a preliminary synthesis with the title *An Attempt to Elucidate the Cosmic Role of Nebulae*, to which we will return later. With this publication, Winnecke's project had come to an end. It is interesting to see that the *Prix Lalande* of the French Académie des Sciences was awarded jointly to H. Kobold and C. Wirtz in 1912, honoring Winnecke's project "to determine with precision the positions of nebulae, and to derive their proper motions, in order to provide information on their distances from the sun", and its execution by the two astronomers (Anonymous 1912). Wirtz was the one who completed Winnecke's work and had gained insight in its implications. So what would he do with it?

2. Wirtz: His Life

Why do we attempt to discuss Wirtz' life and work in detail? First, because he was certainly the most active observer and the most prolific writer among the Strasbourg astronomers. Second, because we can study how World War I interfered with his work, and put it on a different, and also more interesting track. And third, because we can see how the ensuing general political situation between the World Wars influenced his life and his science.

His first published astronomical observation dates from April 1893, when he was $16\frac{1}{2}$ years old. His last paper, a short note about a research trip on a steamer, appeared in January 1939, one month before his death at the age of $62\frac{1}{2}$. His literary output had a "modern" extent: he published about 165 papers, including books and extensive contributions to books, and excluding reviews of books and articles (which sometimes contain quite an amount of additional investigations and discussions on his side), as well as popular accounts of articles, authored by himself or by others, in amateur journals. He studied solar and lunar eclipses, lunar surface features, planets, moons, comets, stars and nebulae; he worked in astrometry, descriptive astronomy, and, in later years, statistical analysis of data; he also worked on position-finding on land and at sea, on atmospheric optics, and on theoretical photometric studies of planetary disks.

It is difficult to classify his work, and to assign it to a given field; he appears to have been a natural philosopher, who used a small telescope (as a student, as a soldier stationed in some place), a fairly large telescope (when he was an observer in Strasbourg), or the meridian circle (when he was working in Kiel), in addition he used pyrheliometers and other types of photometers to carry out atmospheric studies, when he was in Kiel, or was vacationing at the North Sea coast or in the Eifel mountains. In spite of the fact that only a few official files and letters, and and practically no personal notes seem to have survived, the record of his publications and letters helps to reconstruct his life and wanderings. It was the life of a diligent and talented researcher – not that of a genius, but of a full-blooded scientist who did not waste any of his observations, nor any of his colleagues, whose unfinished studies he put into print.

Wirtz was born in Krefeld, Rhineland, on 24 August 1876. His father was the merchant Wilhelm Heinrich Wirtz, his mother was Anna Catharina née Feeger. He visited the secondary school [Gymnasium] in Münstereifel and Krefeld. Before he passed the final secondary school examination [Reifezeugnis] at Krefeld Gymnasium [today's Arndt-Gymnasium] on Easter 1895, he had already carried out color observations of α UMa, and brightness estimates of Jupiter's four Galilean moons, some with a schoolfellow. His final examination record states "he leaves school to study mathematics and natural sciences". In April 1895, he entered Bonn University to study astronomy and mathematics. His astronomy teachers were Friedrich Küstner (1856-1936) and Karl Mönnichmeyer (1860-1942), his teachers in mathematics Hermann Kortum (1836-1904), Rudolf Lipschitz (1832-1903), Ludwig Schlesinger (1864-1933), and Eduard Study (1862-1930), in physics Heinrich Kayser (1853-1940), in geodesy Karl Reinhertz (1859-1906), and in philosophy Johannes Bender (1865-1910).

He obtained his PhD in November 1898. His thesis dealt with the declination measurement of 487 stars (Wirtz 1898). Afterwards he stayed for another year in Bonn to carry out observations of the Hyades with Bonn heliometer between September 1898 and February 1899. From April 1899 to March 1900, he worked at the private Kuffner Observatory[1] in Wien-

[1]The observatory was founded in 1884 by the factory owner Moriz von Kuffner and is a popular observatory today. During Wirtz' stay, Leo de Ball was its director. For details, see Weiss (1985).

Figure 1. A photo of Wirtz, taken before 1904, at the beginning of his time as an observer in Strasbourg. (from Porträtgallerie der Astronomischen Gesellschaft, Stockholm 1904)

Ottakring, doing research on photographic photometry, and continuing his work on the Hyades triangulation. His colleagues were, among others, Karl Schwarzschild and Samuel Oppenheim. In a letter to his colleague Schwarzschild, dated 8 January 1900, he writes: "It is not clear to me what I will do when leaving astronomy; perhaps I will become secondary school teacher, or teacher at a navigation school, perhaps I will succeed to enter the ministry of the navy or of culture in Berlin as a technical assistant..." The Kuffner annual report tells us that "he leaves Vienna in order to become a teacher". He spent the remaining months of 1900 in Bonn to complete and to publish the results of his Vienna work, and to pass the examination as a secondary school teacher [Oberlehrer] for mathematics, physics and geography.

From 1901 to early 1902, he was a teacher at the navigation school [Seefahrtschule] in Hamburg and, from mid-1901, papers on navigation began to appear. Although he mentioned in a letter to Schwarzschild, dated 14 March 1901, that "I don't want ho have anything to do with pure astronomy that has taken away at least three years from my career...", and that he was making progress in the teaching of navigation and was pleasing his director, he achieved an "astronomical comeback": In April 1902, Wirtz (Fig. 1) was appointed observer [Observator] at the Imperial University Observatory in Strasbourg. Was it Küstner's recommendation to Director Ernst Becker, who was looking for a replacement of Hermann Kobold, who had accepted a professorship at Kiel Observatory? Küstner was one of the first graduates of Strasbourg Observatory, and he certainly knew what was needed there to accomplish Winnecke's project. On the other hand, Becker often had problems dealing with his subordinates, and Wirtz, who was a Rhinelander like Becker, was presumably easier to get on with. In any case, it was a good choice. Wirtz' first task was to complete and publish the Bonn heliometer results (Wirtz 1899, 1902).

At the end of the following year, he became private lecturer by having his habilitation accepted. In 1905, he married Helene Borchardt, the daughter of a Berlin banker who had earned his fortune by tea-trade. It should be noted here that Hans Rosenberg, who obtained his PhD in 1905 at Strasbourg Observatory, had married her sister Verena Borchardt a year before. Among the seven brothers and sisters was also the poet Rudolf Borchardt (1877-1945). Wirtz was appointed professor ["Titularprofessor"] in 1909, around the time when Ernst Becker retired from the position of director.

Wirtz' last major observing campaigns in Strasbourg ended in mid-1910. Afterwards, he started a program of photometric observations of nebulae with the large refractor, which would occupy him till his departure from Strasbourg. A son, Daniel Wirtz, was born in 1914, who later studied geology and became professor of geology at Hamburg University in 1954, and who died in 1965. Drafted to the army, Wirtz left Strasbourg on 1 September 1916. After a basic training, he worked at the war headquarters in Berlin in geodesy, astronomy, cartography and ballistics, his duties included the teaching of courses. Nevertheless, astronomical observations were not neglected: in a paper on micrometric observations of a lunar eclipse from Charlottenburg [part of Berlin], completed "in the service of the army", he writes "when carrying out these observations, I enjoyed the assistance of Miss U. Maes, for which I am grateful" (Wirtz 1917a). Let us assume that she was the landlady's little daughter who assisted in writing down some numbers. His geodesy courses resulted in the publication of the book, *Tables and Formulae from Astronomy and Geodesy* (Wirtz 1918a).

His type of publication underwent changes. After finishing his standard observing in 1910, he started editing works by his colleagues Hermann Kobold (who had moved to Kiel University Observatory), Walter Wislicenus (who had died prematurely), and Martin Matzdorff (who had been killed during the heavy fighting at Ieper [Yper], Belgium, in early November 1914). Wirtz undertook orbit calculations of comets and started to evaluate in more details the proper motions of nebulae. Other statistical studies dealt with the stellar magnitudes of the Draper catalogue, color observations of stars, etc.

In Berlin, he made the acquaintance of Paul Hügeler, a bank clerk and editor of the popular astronomical journal *Sirius*, and around early 1917 he started writing articles for it: book reviews and reports on current research of his own studies and those by others, *e.g.* on Einstein's (1917) famous *Cosmological Considerations* (Wirtz 1918b). He also collaborated with Hügeler on some papers of general interest, especially in a study on stellar proper motions, based on Max Wolf's Heidelberg observations, which was submitted in mid-1918, and was published in the Proceedings of the *Heidelberg Academy of Sciences*. During the printing process, Wirtz was absent from Berlin to carry out military-geodetic work in Bohemia, as can be gathered from his correspondence with Max Wolf.

Wirtz' collaboration and literary activity for *Sirius* ended when Hügeler resigned from its editorship in May 1921. Nevertheless, Hügeler had been the one who brought him into closer contact with amateur astronomers, since in 1921 his contributions for the *Astronomisches Handbuch* on solar-system objects, stellar colors, double stars, nebulae and clusters, and atmospheric optics appeared (Wirtz 1921), and in 1922 his brochure *Star Clusters, Nebular Spots and Space* was published, in which many of the "extragalactic" papers of subsequent years were anticipated (Wirtz 1922a).

But we have to step back in time.

After Wirtz' release from the army on 30 January 1919, he could not return to Strasbourg, because it had been taken by the French. So he stayed for a few months in Tübingen, the home of the family of his wife and his brother-in-law, astronomer Hans Rosenberg. An application for the post of director of Königsberg Observatory failed. He found a new employment as observer and non-budgetary [außerplanmäßiger] professor at Kiel University Observatory. A letter of its director, Paul Harzer, reveals that among the 19 applicants for the post of observer and assistant, he was the most excellent. While the person who was best qualified for the post of assistant had accepted another position; it was then filled by Johannes Hellerich. On 1 April 1919, Wirtz started his work in Kiel. This meant that he participated in the "official" program, observations with the meridian circle. As time progressed, it became more and more obvious that the data derived from it were of inferior quality, and Wirtz was certainly happy to dedicate again some of his time for his nebular studies. The first step was a reduction and interpretation of his Strasbourg nebular photometry (Wirtz 1923, 1924a), the second one a statistical investigation of physical properties of nebulae and clusters, which was published in a series of papers in the *Astronomische Nachrichten* and that we shall discuss later.

After the retirement of the Observatory's director Harzer in 1924, Wirtz applied for the position, but various political obstacles prevented the Faculty to accept his application – he appeared to have been not conservative enough. It is possible that Wirtz, around 1920, made the acquaintance of Gustav Radbruch, law professor in Kiel, who was also lecturing, like Wirtz, at the people's university [Volkshochschule] in Kiel. Radbruch was also deputy in the Reichstag for the Social Democratic Party (SPD) and served as Minister of Justice in 1921 and 1923. Wirtz also became a member of the SPD between 1920 and 1930.

This political background, as well as his somewhat ironic humor, meant that *one half of the faculty is against me, the other half is in favor of me, indifferent is no one* (to quote from a later letter by Wirtz), and he was not considered for the directorship. His brother-in-law, Hans Rosenberg, astrophysicist in Tübingen, whose more pronounced nationalistic standpoint, – not at all uncommon among Germans of Jewish origin – was obviously better received by the majority of the faculty, was appointed to the position in 1926. After the rise of Nazi power, Rosenberg soon faced difficulties, which were only temporarily softened because of his status as a front-line fighter [Frontkämpfer] during the World War I. In 1934, he moved to Chicago for a sabbatical year and had to resign his Kiel position in 1935. He remained at Yerkes Observatory till 1937 and, in 1938, became professor and director of the Observatory in Istanbul, Turkey, where he died in 1940.

People at the Ministry of Education [Reichserziehungsministerium (REM)], especially Wilhelm Führer, a former astronomy student from Kiel who had reached a high rank there, were eager to close down Kiel Observatory. The positions were transferred to other observatories in Prussia (first Wroclaw [Breslau], then Poznan [Posen]), as were the instruments. The library was kept in Kiel, and the theoretical physicist Albrecht Unsöld was able to incorporate the library into his institute and to expand it into the "Institut für theoretische Physik und Sternwarte" of later fame by continuing to develop methods for the analysis of stellar atmospheres and to start radio-astronomical work.

In 1937, Wirtz lost his teaching licence: his marriage to a Jewish wife (although being then already a long time at a breakup-point) and his very liberal views were the main reasons. Discharged from the university, he left Kiel and settled with relatives in Krefeld. Since he had been active in studies of atmospheric physics for some years, he carried out research for the

Figure 2. The steamer Buenos Aires of the shipping company Hamburg-Süd. (courtesy of Schiffsbuchhandlung Wolfgang Fuchs, Hamburg)

marine observatory [Seewarte] in Hamburg, including a sea trip on board the steamer Buenos Aires (Fig. 2) from Hamburg to Argentina in Summer 1938. During his sojourn in Argentina, he established some contacts and was offered a post. He also had prospects to find one for his son. However, after his return (Fig. 3), while working on his material at the Seewarte, his health failed, and he died in Hamburg on 18 February 1939.

That he had fallen out of grace among astronomers in Germany is obvious; the *Vierteljahrsschrift der Astronomischen Gesellschaft* (for which he had contributed numerous reviews) did not mention his death. Who would have been able to write it? His brother-in-law Rosenberg, who had escaped to Istanbul? No one would have published it. Other former colleagues at Kiel Observatory were deeply engaged in the ruling system, Joachim Stobbe, the astronomer who had taken over Wirtz' position in Kiel, had become professor in Berlin-Babelsberg, while his colleague Kobold had already denounced him at the time when the director's position had to be filled again. They would certainly not have liked to tarnish their "reputation" by writing an obituary for a dissident. The *Astronomische Nachrichten* reported his death in a single line. Only in the amateur journal *Himmelswelt*, edited

Figure 3. Carl Wirtz on board the steamer Buenos Aires, summer 1938. (courtesy of Renate and Bettina Wirtz).

by the liberal astronomer Hans Kienle with whom Wirtz had emptied some bottles of wine on the occasion of a total solar eclipse expedition in Lapland in 1928, a 12-line obituary appeared: "One of the best-known German astronomers, whose scientific studies in diverse fields, carried out with a wilful and critical mind, have given him a special reputation in the circles

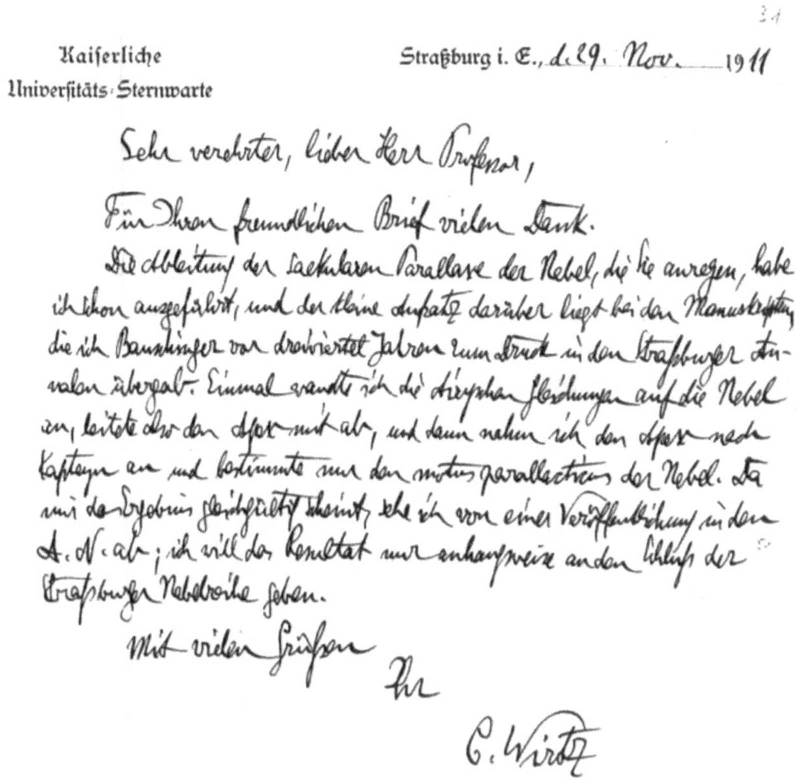
Kaiserliche
Universitäts-Sternwarte

Straßburg i. E., d. 29. Nov. 1911

Sehr verehrter, lieber Herr Professor,

Für Ihren freundlichen Brief vielen Dank.

Die Ableitung der saekularen Parallaxe der Nebel, die Sie anregen, habe ich schon ausgeführt, und der kleine Aufsatz darüber liegt bei den Manuskripten, die ich Bauschinger vor dreiviertel Jahren zum Druck in den Straßburger Annalen übergab. Einmal wandte ich die Airyschen Gleichungen auf die Nebel an, leitete also den Apex mit ab, und dann nahm ich den Apex nach Kapteyn an und bestimmte nur den motus parallacticus der Nebel. Da mir das Ergebnis gleichgültig scheint, sehe ich von einer Veröffentlichung in den A. N. ab; ich will das Resultat nur anhangsweise an den Schluß der Straßburger Nebelreihe geben.

Mit vielen Grüßen

Ihr

C. Wirtz

Figure 4. Letter from Carl Wirtz to Karl Schwarzschild on his work on the motion of nebulae in 1911. (courtesy of Staats- und Universitätsbibliothek Göttingen)

of his colleagues" (Anonymous 1939).

3. Wirtz: His Scientific Activity in Strasbourg

As already mentioned, Wirtz' activities in Strasbourg had been manifold. He was not only appointed as an "observer", he truly was an observer, his activities covered the observation of solar and lunar eclipses, planets, satellites, comets, double and variable stars. He even wrote papers on the variability of lunar craters, and on the figure of the moon. Besides this, he carried the main workload of reducing the nebular observations accumulated by his predecessors and colleagues at Strasbourg. When most of these tasks were completed, he started a work of nebular photometry that was, however, cut short by the war, and only completed and published later (Wirtz 1923).

Wirtz' major achievement in "nebular" astronomy during his Strasbourg time is the reduction of the observations, the establishment of the general catalog of nebular spots (1881-1910) and its comparison with other lists of nebulae, as published in the first part of the 4th volume of the Strasbourg *Annalen* (1911). The second part of the 4th volume (1912) contains, besides observations of solar-system objects and double stars, a short paper *Addenda to the Work on Nebular Spots, including an Attempt to Elucidate the Cosmic Role of Nebulae*, that he also mentioned in his last letter to Schwarzschild, written on 29 November 1911[2] (Fig. 4). Here he tried to combine the Strasbourg visual positions and Heidelberg positions, derived by Lorenz (1911) from photographs, and to compare them with the earlier list of visual positions by H. Schultz (1874) in Uppsala, to find out whether there was a systematic trend in the differences of the nebular positions determined at different times.

Wirtz (1912) found: "that the nebular apex derived both for the Strasbourg – Uppsala and the Heidelberg – Uppsala dataset was noticeably different from the stellar apex", *i.e.* stars and nebulae seem to belong to two completely different systems.

An average proper motion of the objects away from the solar apex can also be converted into a secular parallax, which is an indicator for the average distance of a group of objects. Wirtz derived a mean parallax $\pi = -0.0020'' \pm 0.0024$, which indicated that the nebulae must be very distant objects, and he concluded "in the present stage, a decision cannot be made as to whether the computed nebular drift is real or accidental."

4. Wirtz: His Scientific Activity During the War

It took several years until Wirtz picked up the topic again. In 1916, he continued with two more papers on *The drift of the nebular spots.* Wirtz' (1916a&b) conclusions were:

"1. The large mass of nebulae drifts towards an apparent vertex in RA$= 245°$, Decl $= -3°$. The apparent direction of the dominant nebular motion is almost opposite to the parallactic motion of the fixed stars.

[2]The translation of the text in Fig. 4 runs as follows:
Very estimated, dear Professor,
Thank you very much for your kind letter.
The derivation of the secular parallax of nebulae that you suggest I have already carried out, and the little note is with the manuscripts that I gave to Bauschinger three quarters of a year ago to be published in the Strasbourg Annals. First, I applied Airy's equations to the nebulae, thus I derived the apex, too, and then I assumed the apex according to Kapteyn and only derived the *motus parallacticus* on the nebulae. Since the result appears to be of no value to me, I refrain from publishing it in the *Astronomische Nachrichten*; I intend to give the result only as an appendix to the end of the Strasbourg list of nebulae.
With many regards,
Yours C. Wirtz.

2. The gaseous nebulae have the same parallactic motion as the fixed stars, they are located at the limit of the solar star cluster which surrounds our position in space."

In modern words, the gaseous nebulae (mainly the planetary nebulae) behave like distant stars, while the extragalactic nebulae behave differently (and their motion rather reveals the deficiencies of the reference system).

Objections against Wirtz' conclusions were raised by Hermann Kobold (1916), his former colleague of Strasbourg, who was now in Kiel, as well as by the dean of German astronomy, Hugo von Seeliger (1916) in Munich. Wirtz (1917) replied to this critique in a paper where he suggested a new idea:

"Should the small nebulae have to be considered at rest with respect to the system of fixed stars, we would by this gain a valuable hint; because the small objects which can be centered reliably on the photographic plates would yield a new reference system and zero points for the studies of stellar motions, as would be apparent even from the few, not very accurate micrometer measurements."

Only 70 years later, in 1987, a solar apex determination based on an extragalactic reference frame was published!

The final paper of Wirtz during the war years, *On the Motion of the Nebular Spots* (Wirtz 1918c), was concerned with proper motions and radial velocities. Here he spoke even more clearly about the stationarity of the nebulae on the sphere, "one would have, with the nebulae, an excellent reference system for stellar motions". Radial velocities compiled by Paddock (1916) – but mainly determined by V.M. Slipher at Lowell Observatory – had reached Wirtz by December 1917, in spite of the USA having entered the war. Unlike Paddock, who despised the huge k-term in his solution, Wirtz believed that it was a real effect. His conclusions were: "It is remarkable that our system of fixed stars should have such an incredibly large displacement of 820 km/sec, and equally strange is the interpretation of the systematic constant $k = +656$ km/sec." and he added, contradicting Paddock, "Experience shows again and again that the law of large numbers emerges already at remarkably small quantities of things."

5. Wirtz: His Scientific Activity in Kiel

Several papers on nebulae were written by Wirtz in his early Kiel days. We already mentioned his writings for amateur astronomers, who somehow helped to focus his view on the interpretation of his old and of newly published observations.

His paper on *Some Topics Concerning the Statistics of Spiral Nebulae and Globular Clusters* (Wirtz 1922b) already highlights the way ahead:

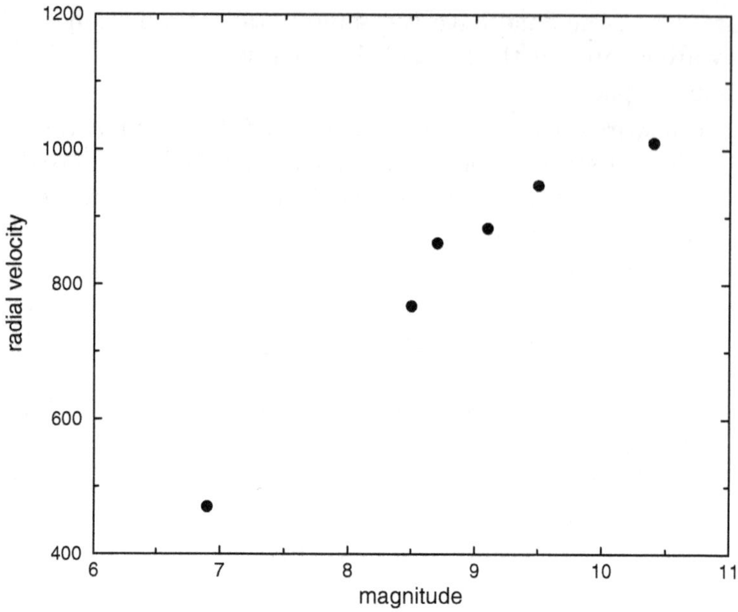

Figure 5. The magnitude-radial velocity relation of spiral nebulae by Wirtz (1922b), reconstructed from his data.

"At present the radial motions of 29 spiral nebulae are available in the literature. Though it is definitively expected that this material, being as scarce as it is valuable, will gradually grow, it does not seem premature to search for rules, even in these few data, which point to systematic interrelations. In other ways, and from different kinds of material, common properties have already been recognized and have been worked into a hypothesis of the structure of the universe."

He searched for correlations between radial motions and other properties. He found a weak one with galactic latitude and a strong one with magnitude, which he presented as a table, and which is shown as a reconstructed graph in Fig. 5. He wrote:

"Very pronounced is the parallelity of v with the total magnitude Mg. In the mean values [...] obtained [...] an approximately linear relation is present in the sense that the spiral nebulae nearer to us have a tendency of approaching, the more distant ones a tendency of receding from our Milky Way system... One finds, of course, analogous behaviour also in the diameters of the structures, where a positive change of distance belongs to small diameters, and a negative one to large diameters.

All these statistical features superimpose on the most striking major process, which can be described as an expansion if the system of spiral nebulae with respect to our own position... and the dependence on mag-

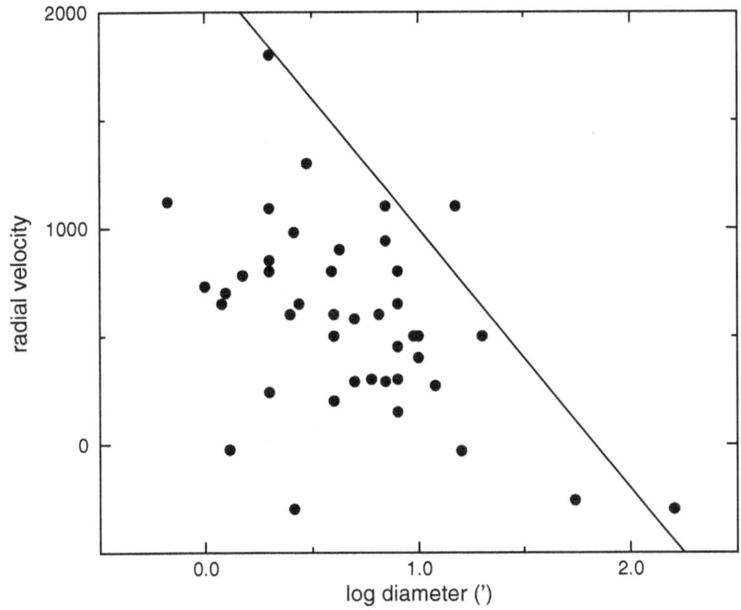

Figure 6. The diameter-radial velocity relation of spiral nebulae by Wirtz (1924b), reconstructed from his data.

nitude indicates that the nearer or the most massive nebulae show less expansion than the distant nebulae or those of lesser mass."

The last step of Wirtz was to find a theoretical backing of his observations, and this he described in his last paper *De Sitter's Cosmology and the Radial Motion of the Spiral Nebulae* (Wirtz 1924b). He took his theoretical insight mainly from Eddington (1923) and Weyl (1924): first the general scattering of matter towards the mass horizon, and then the spectral line shift towards the red which appears in distant objects even when they are at rest with respect to the origin of coordinates, due to the slower course of time.

Taking the redshift data from Eddington, and nebular diameters from various catalogues, Wirtz investigated the correlations between the measured quantities. He wrote:

"Though nothing is known about the distances of the spiral nebulae, under the assumption that on the average the linear diameters are equal, one can use their apparent diameters as a measure of distance. In de Sitter's cosmology, the radial motion should increase with decreasing diameter."

Wirtz constructed a graph, which he described as follows:

"When the different quantities v versus $\log Dm$ are represented graphically, the diagram shows a V-shaped or triangular form from which one deduces the following facts: for the apparent small nebulae one finds small-

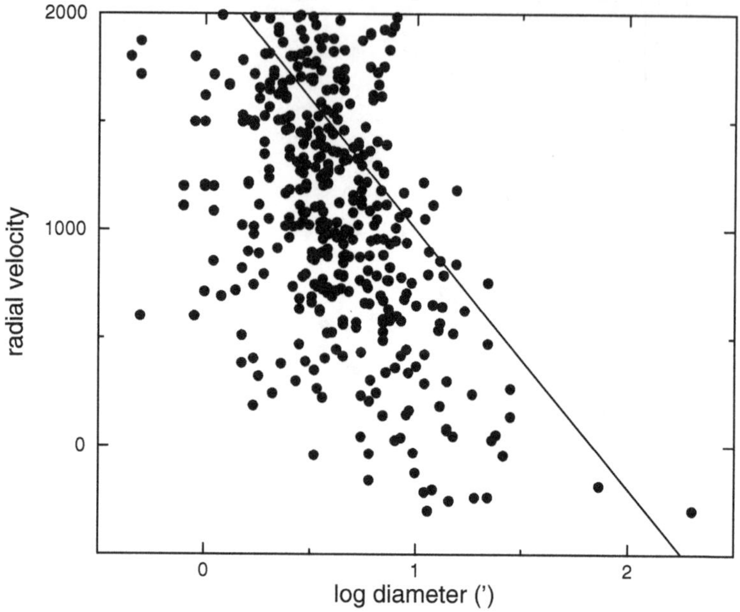

Figure 7. The diameter-radial velocity relation of spiral nebulae by Wirtz (1924b), shown with modern data (UGC).

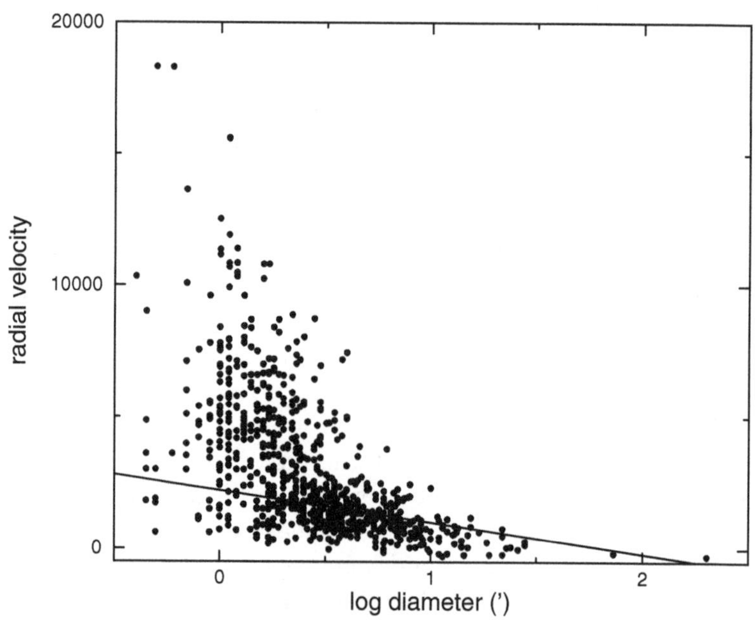

Figure 8. The diameter-radial velocity relation of spiral nebulae by Wirtz (1924b), shown with modern data (UGC).

est and largest v, those with apparently largest values have the smallest v, among the nebulae with small v one finds large and small objects, large nebulae with large v do not exist.

From this, one deduces that the dispersion of the linear dimensions of the nebulae fills the triangular plane in such a way that among the near nebulae absolutely small and large objects are visible while in the depth of space only the absolutely largest are subject to observing their radial motions. The progression between v and apparent Dm is reproduced best when the hypotenuse of the triangle enclosing all observed points, which follows the absolutely largest nebulae under the assumption that the giants among the nebulae have the same average extent at all distances."

The diagram, reconstructed from Wirtz' data, is shown in Fig. 6. He found:

"For the [absolutely] largest nebulae the graphic presentation then yields the formula

$$v(km) = 2200 - 1200 \log Dm$$

(Dm in arc minutes). The coefficient 1200 indicates that, with a tenfold increase of distance, the redshift computed as pure Doppler effect increases by 1200 km. This rule of thumb has the advantage that it does not come into conflict with the velocity of light, because the real and apparent radial motions which result from the de Sitter's space then increase only very slowly with distance; and a redshift from which, in terms of the Doppler effect, the velocity of light follows, would be reached only at distances which lie far beyond the opinions now held about the positions of the celestial bodies – approximately 10^{200} parsec, a quantity which exceeds also all estimates of the radius of curved space."

Besides the fact that his theoretical interpretation is somewhat flawed, he was obviously the first (but certainly not the last) observational cosmologist who was led astray by selection effects. If we use modern data from the *Uppsala General Catalogue* (Nilson 1973) within the limits of his diagram, the relation still looks reasonable (Fig. 7), but if we include all suitable UGC data, we see that he found the correct trend, but not the correct relation (Fig. 8).

The year 1924 saw several constructions of velocity-distance relations that were intended to prove the existence of the "de Sitter effect" (de Sitter 1917) in astronomical data: by Wirtz, Silberstein, and Lundmark. Actually Silberstein (1924) was the first one to publish, but his focus was on globular clusters which are unsuited for this kind of study, something that Silberstein never realized in his life. Wirtz (1924) was second, and the first person to use the spiral nebulae. Lundmark (1924), who was using own distance determinations of nebulae, was third. Later attempts were made by Strömberg (1925), Wirtz' PhD student Dose (1927), Lemaître (1927),

Robertson (1928) and Hubble (1929), but only Lemaître's work (because of its modern theoretical background), and Hubble's work (because of its improved nebular distances) have received due attention. Both authors were first to obtain a linear velocity distance relation – Lemaître from his theory, and Hubble from his careful observations.

6. Wirtz and the Extragalactic Nebulae: An Appraisal

As we have seen, Wirtz' first empirical relation was established between velocities and magnitudes. He used six normal points, based on 29 objects. The correlation coefficient for the $v(m)$-relation is $cc = 0.972$. We know that Hubble's linear relation between distance and velocity results in a linear $\log v(m)$ relation, but with Wirtz' data, its cc is only 0.950. Was this, perhaps, the reason why Wirtz chose to plot v versus m?

For a homogeneous expanding universe, the expected slope is $\log z(m) = 0.2$. Wirtz' (1922) normal points yield a slope of 0.098. This provides an early example of a magnitude-limited sample with significant losses of data near the nominal (faint) limit, which later became known as the Malmquist bias (Malmquist 1922). Hubble's (1929) linear diagram shows radial velocity versus distance. The $\log z(m)$-diagram, later known as the Hubble diagram, was first introduced by de Sitter (1930) and next used by Hubble & Humason (1931).

When in 1924 Wirtz used galaxy diameters to test the de Sitter universe, he did it empirically (since no clear theoretical prediction about the shape of the relation was available). His relation of $\log Dm(v)$ is consistent with $m(v)$, the relation he already used in 1922. However, as we have seen, Wirtz' sample was magnitude-limited and thus plagued by the Malmquist bias.

Wirtz continued his extragalactic research when Rosenberg had become director. An early attempt was made to derive the luminosity function of galaxies, using total magnitudes by Fath and angular diameters as distance indicators (Wirtz 1926). This was followed by a more substantial photographic photometry of the Virgo-Coma group (Wirtz 1927), where the first indication of luminosity segregation in a cluster was found.

Like many early workers in cosmology, Wirtz was later overlooked. Friedmann (1922) was despised by Einstein and had been forgotten, Lemaître at least for some years ignored, Lundmark and de Sitter attacked and despised by Hubble (because of their parallel work in nebular classification and the velocity-distance relation), and Wirtz, who had no observational possibilities and hardly any support in his own university and country, was forgotten, too. His colleagues and students dispersed, his manifestation during his last years was that of a lonely professor *giving popular lectures to interested old ladies* (to quote Unsöld who met him when he came to Kiel as a young physics professor). Times had changed, and not necessarily to the better.

To quote from some of his letters written to relatives in late 1935: *I have never been at home in my life. But perhaps the character of the human soul is to only feel as a guest on earth... More and more I have the clear insight that I have missed my homeland of time, rather than my homeland of space. My heart belongs to antiquity, not to the present time* (Köppen 1967). Wirtz' (1936) last reminder on his work on the motion of spirals was actually that of a person who had lost track of current developments. He may have failed in most of his work on the extragalactic nebulae, but he was the first who tried new paths of research: he was attacked by some, ignored by others. Those who followed his tracks had by then forgotten who paved the road.

7. Sources and Acknowledgments

Documents are kept in the Bundesarchiv (Berlin), the Geheimes Staatsarchiv Preußischer Kulturbesitz (Berlin), the Staats- und Universitätsbibliothek Göttingen (letters to Schwarzschild), and the Universitätsbibliothek Heidelberg (letters to Wolf and Radbruch).

We thank the Bundesarchiv (Berlin), the Geheimes Staatsarchiv Preußischer Kulturbesitz (Berlin) and the former DDR Staatsarchiv (Merseburg) for granting access to the documents, and the libraries in Göttingen and Heidelberg for sending copies of the letters. We furthermore thank the university archives of Bonn and Muenster for providing additional information.

We also cordially thank the following persons who provided us with material and information: Renate and Bettina Wirtz (Hannover), Wolfgang Fuchs (Hamburg), Albrecht Gierschmann (Krefeld), Felix Lühning (Bremen), and Charlotte Schönbeck (Wilhelmsfeld).

References

1. Anonymous 1912, Prix Lalande, *C.R. Acad. Sc. Paris* **155**, 1302.
2. Anonymous 1939, Umschau, *Himmelswelt* **49**, 147.
3. Becker, E. 1909, Beobachtungen von Nebelflecken ausgeführt in den Jahren 1875 bis 1880 am 6-zölligen Refraktor der provisorischen Sternwarte von A. Winnecke, *Annalen der Kaiserlichen Universitäts-Sternwarte in Strassburg*, Band 3, Teil 1.
4. de Sitter, W. 1917, On Einstein's Theory of Gravitation, and its Astronomical Consequences – Third Paper, *Monthly Not. R. Astron. Soc.* **78**, 3.
5. de Sitter, W. 1930, On the Magnitudes, Diameters, and Distances of the Extragalactic Nebulae, and their Apparent Radial Velocities, *Bull. Astron. Inst. Netherlands* **5**, 157.
6. Dose, A. 1927, Zur Statistik der nichtgalaktischen Nebel auf Grund der Königstuhl-Nebellisten, *Astron. Nachr.* **229**, 157.
7. Eddington, A.S. 1923, The mathematical theory of relativity, Cambridge University Press, Cambridge.

8. Einstein, A. 1917, Kosmologische Betrachtungen zur allgemeinen Relativitätstheorie, Sitzungsberichte Berlin 1917.
9. Fath, E.A. 1909, The Spectra of Some Spiral Nebulae and Globular Star Clusters, *Lick Obs. Bull.* **5**, 71.
10. Friedmann, A. 1922, Über die Krümmung des Raumes, *Zs. Physik* **10**, 377.
11. Hubble, E. 1929, A Relation between Distance and Radial Velocity among Extragalactic Nebulae, *Proc. Natl Acad. Sci. Washington* **15**, 168.
12. Hubble, E. & Humason, M.L. 1931, The Velocity-Distance Relation among Extragalactic Nebulae, *Astrophys. J.* **74**, 43.
13. Huggins, W. & Huggins, M.L.M. 1899, An Atlas of Representative Stellar Spectra from 4870 to 3300, W. Wesley and Son, London.
14. Kobold, H. 1916, Bemerkung zur Untersuchung der Eigenbewegungen der Nebelflecke, *Astron Nachr.* **203**, 299.
15. Köppen, E. 1967, Krefelder Miniaturen, Scherpe, Krefeld, p. 265.
16. Lemaître, G. 1927, Un univers homogène de masse constante et de rayon croissant, rendant compte de la vitesse radiale des nébuleuses extra-galacticques, *Ann. Soc. Scient. Bruxelles* **47**, 494.
17. Lorenz, W. 1911, Photographische Positionsbestimmungen von 178 Nebelflecken, *Veröff. Grossh. Sternw. Heidelberg (Königstuhl)* **6**, No. 4, 19.
18. Lundmark, K. 1924, The Determination of the Curvature of Space-time in de Sitter's World, *Monthly Not. R. Astron. Soc.* **84**, 747.
19. Malmquist, K.G. 1922, On Some Relations in Stellar Statistics, *Arkiv Mat. Astron. Fys.* **16**, No. 23.
20. Nilson, P. 1973, The Uppsala General Catalogue of Galaxies, *Uppsala Astron. Obs. Annaler* **6** (UCG).
21. Paddock, G.F. 1916, The Relation of the System of Stars to the Nebulae, *Publ. Astron. Soc. Pacific* **28**, 109.
22. Robertson, H.P. 1928, On Relativistic Cosmology, *Phil. Mag. (Ser. 7)* **5**, 835.
23. Scheiner, J. 1899, Ueber das Spectrum des Andromedanebels, *Astron. Nachr.* **148**, 325 (also *Astrophys. J.* **9**, 149).
24. Schultz, H. 1874, Micrometrical Observations of 500 Nebulae, *Nova Acta Societas Scientiarum Upsalensis*, Ser. 3, 9, vol. 2.
25. Seeliger, H. v. 1916, Über die Eigenbewegungen der Nebelflecke, *Astron. Nachr.* **203**, 305.
26. Silberstein, L. 1924, The Curvature of de Sitter's Space-time Derived from Globular Clusters, *Monthly Not. R. Astron. Soc.* **84**, 363.
27. Strömberg, G. 1925, Analysis of Radial Velocities of Globular Clusters and Non-galactic Nebulae, *Astrophys. J.* **61**, 353.
28. Weiss, W.W. 1985, Die Kuffner-Sternwarte, Jugend und Volk-Verlagsgesellschaft, Wien.
29. Weyl, H. 1924, Massenträgheit und Kosmos, *Naturwissenschaften* **12**, 197.
30. Winnecke, A. 1875, Ueber die... Beobachtungen der Nebelflecke, *Vierteljahrsschrift Astron. Ges.* **10**, 297.
31. Wirtz, C.W. 1898, Bestimmung der Deklinationen von 487 Sternen und der Polhöhe der Bonner Sternwarte aus Beobachtungen am vierzölligen Ertel'schen Passageninstrument im Ersten Vertikal, *Veröff. Kgl. Sternw. Bonn* **3**.
32. Wirtz, C.W. 1899, Triangulation der Hyadengruppe, ausgeführt am sechszoelligen Fraunhoferschen Heliometer der Bonner Sternwarte, Manuscript, Bibliothek der Astronomischen Institute der Universität Bonn (Inv.-Nr. 5349, Sect. 31/28).
33. Wirtz, C.W. 1902, Triangulation der Hyadengruppe, ausgeführt in den Jahren 1898-99 am sechstzölligen Fraunhoferschen Heliometer der Bonner Sternwarte, *Astron. Nachr.* **160**, 17.
34. Wirtz, C.W. 1909, Beobachtungen von Nebelflecken ausgeführt in den Jahren 1880-1902 am 18-zölligen Refraktor der neuen Sternwarte von H. Kobold, A. Winnecke

und W. Schur, *Annalen der Kaiserlichen Universitäts-Sternwarte Strassburg* **3**, Teil 2.
35. Wirtz, C.W. 1911a, Beobachtung von Nebelflecken am 49 cm Refraktor 1902 April bis 1910 März, *Annalen der Kaiserlichen Universitäts-Sternwarte Strassburg* **4**, Teil 1, 1.
36. Wirtz, C.W. 1911b, Generalkatalog der am großen Refraktor beobachteten Nebelflecke 1881-1910, *Annalen der Kaiserlichen Universitäts-Sternwarte Strassburg* **4**, Teil 1, 79.
37. Wirtz, C.W. 1911c, Vergleichung des Generalkataloges mit anderen Nebelverzeichnissen, *Annalen der Kaiserlichen Universitäts-Sternwarte Strassburg* **4**, Teil 1, 113.
38. Wirtz, C.W. 1912, Ein Versuch zur kosmischen Stellung der Nebel, *Annalen der Kaiserlichen Universitäts-Sternwarte Strassburg* **4**, Teil 2, 313.
39. Wirtz, C.W. 1916a, Die Trift der Nebelflecke, *Astron. Nachr.* **203**, 197.
40. Wirtz, C.W. 1916b, Die Trift der Nebelflecke – Zweite Mitteilung, *Astron. Nachr.* **203**, 293.
41. Wirtz, C.W. 1917a, Mikrometrische Beobachtungen der Mondfinsternis 1917 Juli 4, *Astron. Nachr.* **205**, 135.
42. Wirtz, C.W. 1917b, Über die Eigenbewegungen der Nebelflecke, *Astron. Nachr.* **204**, 23.
43. Wirtz, C.W. 1918a, Tafeln und Formeln aus Astronomie und Geodäsie für die Hand des Forschungsreisenden, Geographen und Geodäten, Springer, Berlin.
44. Wirtz, C.W. 1918b, Der Weltraum, ein geschlossener sphärischer Raum (sign. W.), *Sirius* **51**, 21.
45. Wirtz, C.W. 1918c, Über die Bewegungen der Nebelflecke. Vierte Mitteilung, *Astron. Nachr.* **206**, 109.
46. Wirtz, C.W. 1921, Various chapters in *Astronomisches Handbuch, herausgegeben vom Bund der Sternfreunde, Stuttgart*, Francksche Verlagshandlung.
47. Wirtz, C.W. 1922a, Sternhaufen, Nebelflecke und Weltraum, Astronomische Schriften des Bundes der Sternfreunde 1. Franck, Stuttgart.
48. Wirtz, C.W. 1922b, Einiges zur Statistik der Radialgeschwindigkeiten von Spiralnebeln und Kugelsternhaufen, *Astron. Nachr.* **215**, 349.
49. Wirtz, C.W. 1923, Flächenhelligkeiten von 566 Nebelflecken und Sternhaufen, *Medd. Lunds Astron. Obs.*, Ser. II, No. 29.
50. Wirtz, C.W. 1924a, Kugelnebel, Spiralnebel und Flächenhelligkeit – Eine statistische Studie. In Probleme der Astronomie, Festschrift für Hugo v. Seeliger, Springer, Berlin, p. 66.
51. Wirtz, C.W. 1924b, De Sitters Kosmologie und die Radialbewegungen der Spiralnebel, *Astron. Nachr.* **222**, 21.
52. Wirtz, C.W. 1926, Totalhelligkeit und Leuchtkraftkurve der nichtgalaktischen Nebel, ein Versuch, *Astron. Nachr.* **228**, 41.
53. Wirtz, C.W. 1927, Zur Photometrie der nichtgalaktischen Nebel – Die Virgo-Coma-Gruppe, *Publ. Sternw. Kiel* **15**, No. 3, 35.
54. Wirtz, C.W. 1936, Ein literarischer Hinweis zur Radialbewegung der Spiralnebel, *Z. Astroph.* **11**, 261.
55. Wolf, M. 1909, Über einige Nebelspektren, *Astron. Nachr.* **180**, 151.

PAUL MULLER (1910-2000)

P. BACCHUS
Université de Lille
40, rue Haute
La Grande Paroisse
F-77130 Montereau, France
pierre.bacchus@worldonline.fr

Abstract. This note is devoted to Paul Muller who started his career at Strasbourg Observatory and became a world-wide specialist on double stars.

Paul Muller entered in 1931 at Strasbourg Observatory, recently directed by André Danjon. The latter entrusted him with studying the birefringence of quartz prisms and of its applications to astronomy. Taken prisonner in 1940, Muller spent five years in an Offlag[1] before returning to Strasbourg. In 1948, he presented his thesis entitled "Sur un nouveau micromètre à double image, ses possibilités et quelques questions connexes"[2] in Paris where Danjon had been nominated as Director at the end of WWII.

From then on, most of his activities were devoted to double stars. His micrometer, installed on the small 16cm equatorial telescope and later on the 49cm Large Refractor, demonstrated a precision and a convenience of usage much superior to those of wire micrometers. Muller adapted the micrometer to the measurement of magnitude differences between components which were until then only estimated. With the help of colored filters, he also measured the color differences – which had never been done – as well as the diameters and the flattening of planets and satellites.

While on a mission at the 91cm telescope of Lick Observatory, Muller's abilities as an observer together with his micrometer's allowed him to mea-

[1]Concentration camp for officers.
[2]On a new double-image micrometer, its abilities and a few related matters.

A. Heck (ed.), The Multinational History of Strasbourg Astronomical Observatory, 189–190.

Figure 1. Paul Muller (right) with Director Pierre Lacroute at the Great Refractor of Strasbourg Observatory in the early 1950s. (Courtesy Mrs. L. Muller)

sure double stars with a separation of only 0.1″ and to start discovering long series of new couples.

In order to make easier and to foster the study of double stars, Muller took two important initiatives: establishing a (then manual) file of all double stars observations and launching in 1954, under the sponsorship of IAU Commussion 26, a periodical circular gathering together news on double stars. He ensured himself, until 1983, the publication of the first ninety issues. He had of course become a member of IAU Commission 26 (Double Stars) which he chaired several times from 1952 on.

Muller left Strasbourg in 1956 to join Meudon Observatory, then Nice Observatory, where the two largest French refractors allowed him to pursue his studies of double stars. But in parallel, and from the launch of the first artificial satellite in October 1959, he tracked their positions and set up the French network of optical observing stations.

His fondness for public outreach was demonstrated at various opportunities and in particular by his participation to the foundation of the Alsace Group of the *Société Astronomique de France (SAF)* in 1931.

VISTAS INTO THE CDS GENESIS

A. HECK
Observatoire Astronomique
11, rue de l'Université
F-67000 Strasbourg, France
heck@astro.u-strasbg.fr

Abstract. This chapter gathers together a few milestones of the pre-CDS years marking the advent of astronomical data centers. Two chronologies are reviewed: one essentially at the level of the International Astronomical Union (IAU), and the other one centered on the preparatory steps to the official creation, in 1972, of the *Centre de Données Stellaires (CDS)* in Strasbourg. Some tentative explanations to CDS' long-term success are put forward.

1. The Advent of Data Centers

The 14th General Assembly (GA) of the International Astronomical Union (IAU) was officially declared open on 18 August 1970 in Brighton (UK) by a freshly appointed (June 1970) Secretary of State for Education and Science who was going to stay in history as the Iron Lady after three terms as British Prime Minister (1979-1990). During her opening speech, Margaret Thatcher stressed the contributions made by government-supported establishments such as the Royal Observatories at Herstmonceux and Edinburgh. The Royal Greenwich Observatory (RGO) located at Herstmonceux actually moved to Cambridge in 1990 before being definitively closed on 31 October 1998.

That 14th IAU GA will however stay in history, for what concerns us here, as the first one where the issue of astronomy-related data centers has been officially debated at specific meetings.

This was prepared by various steps expressing the fact that, towards the end of the 1960s, the idea of data centers (conceptually in the continuation of cataloguing) had been popping up here and there among astronomers

A. Heck (ed.), The Multinational History of Strasbourg Astronomical Observatory, 191–209.

together with individual initiatives setting up what would be seen today as embryonic materializations.

At the level of the IAU, a first document dated June 1969 and directed by the then IAU General Secretary L. Perek to the members of the IAU Executive Committee (EC) can be singled out. As it is an excellent summary of the spirit of the times, this document is reproduced *in extenso* hereafter.

INTERNATIONAL ASTRONOMICAL UNION
Union Astronomique Internationale

EC 32 DOC. 11 a

TO THE MEMBERS OF THE EXECUTIVE COMMITTEE

Prague June 1969

Dear Colleague,

Recently I had the opportunity to discuss some problems of astrophysical data with C. Jaschek, who draw [drew?] my attention to his paper in the PASP 80, 654, 1968. In what follows I tried to formulate the problem on somewhat more general lines.

The problem of making Scientific Information accessible to all users received much attention in recent years. Thus:

- UNESCO[1] and ICSU[2] created a committee for UNISIST[3] – a World Science and Information System,
- ICSU created a special committee CODATA – Committee on Data for Science and Technology which is very active in the field of spectroscopy (our representative in CODATA is Mrs. Moore-Sitterly, former President of Commission 14),
- the computing of data for astronomical yearbooks had successfully been coordinated and resulted in an important saving of efforts,
- astronomical bibliography is now handled by modern methods in Astronomy Abstracts[4] which will start appearing soon,
- etc.

As an intermezzo, imagine an astronomer who is interested in a rather faint star and would like to know if it has an MK classification. He would look through 5, 10, 100 papers, according to his temperament, but would certainly give up his search before half through the relevant literature. He would then probably travel to the nearest observatory with a good spectrograph, take the plate and determine the type. Or, he would get interested in another project

[1]United Nations Educational, Scientific and Cultural Organization.
[2]International Council of Scientific Unions.
[3]Universal System for Information in Science and Technology.
[4]Renamed as *Astronomy and Astrophysics Abstracts*, the continuation of the *Astronmischer Jahresbericht* launched in 1899 by W.F. Wislicenus in Strasbourg (see Duerbeck 2005).

and shelve that particular star. This situation is evidently undesirable and unfortunately it is not limited to the MK spectral classes. It concerns the whole field of data resulting from astrophysical observations.

We could imagine other painful situations such as that of two astronomers observing the same object under different names[5] ... but let this be left to your phantasy.

It may be concluded that information scattered over a large number of papers is for all practical purposes inaccessible to individual scientists. Unless the data are compiled in a systematic and homogeneous way, the compilation is kept up to date, and the dissemination made easy, the observed data cannot be used.

The areas with which we should be concerned are in particular:

- radial velocities
- spectral types
- rotational velocities
- photoelectric magnitudes
- orbits of visual binaries
- orbits of spectroscopic binaries
- catalogues of bright stars
  binaries
  variable stars
  planetary nebulae
  diffuse nebulae
  star clusters
  galaxies
  radio sources
  X-ray sources
  quasistellar sources
- etc.

Some of the above lists or catalogues do exist, were recently published and are excellent from all points of view, but we cannot be sure that:

1. All necessary catalogues really do exist,
2. The catalogues are reissued at proper intervals, which may differ for each of the above items,
3. The catalogues are organized in the best possible way,
4. The material in all areas, where it is desirable, is compiled at one and only one place, that there are no gaps and no duplications,
5. All objects, in particular of newly discovered classes, are designated by the same names or symbols by all astronomers.

In the past the IAU contributed in a substantial way to coordination. The names of variable stars have been accepted by everybody, and the same applies to the lunar nomenclature of the near side (and we hope that that of the far side will be universally accepted, too). It is not necessary to enumerate all the international projects where the IAU played an important role of coordinating. I feel that the IAU should assist the astronomical community also in this case.

[5]This actually happened without the two authors realizing they were talking of the same object (a star in that case).

In my personal views, the situation could be improved either by setting up an ad hoc committee or by entrusting the task of coordination to Commission 5. The task would consist of:

1. Convening all those interested in the problem during the General Assembly in Brighton,
2. Stimulating the interest of Commissions,
3. Establishing general principles for the collection and dissemination of data.

All further actions should depend on the Commissions, in particular:

1. Deciding what material should be compiled and in which way it should be kept and disseminated,
2. Arousing interest of institutions and individuals who are in a position to act as data centres,
3. Urging individual observers to transmit their data to the centres,
4. Naming newly discovered objects,
5. Investigating possibilities of printing catalogues or lists at convenient intervals.

May I have your comments, either in writing or at the Executive Committee meeting?

L. Perek
General Secretary

One remains impressed by the clearsightedness of the detailed problematics, at a time it was not yet fully nor acutely emphasized by the *massification* of data and by their recording plus storage in machine-readable formats.

The paper by Jaschek (1968) mentioned in the opening paragraph of that IAU document discusses some aspects of the collection and cataloguing of observational data. It recommends several data collection centers for storing information and disseminating it at convenient intervals to the scientific community. Jaschek, who will become in 1974 the second CDS Director, brings up a number of interesting considerations such as:

> "The principal advantage offered by [a data center] would be the fact the cataloguing would be done as a full-time job on a permanent and secure basis"
> [...]
> "critical evaluations [...] require professional astronomers to do the job".

The paper ends as follows:

> "... it seems to the writer that the data center provides the only sensible answer to the necessity of improving the accessibility of published data. The idea of a data center – either one or several – has been discussed recently on at least two occasions; namely the Commission Meeting of IAU Commission No. 45 at

Prague, under the presidency of W.P. Bidelman, and the conference sponsored by the National Science Foundation, on the "Construction and Use of Star Catalogues" held in 1966 at the University of Maryland."

(Jaschek 1968, p. 660)

Looking in turn at those two references, we see in the proceedings of the IAU Commission 45 meeting at the 13th IAU GA (Prague, 1967) that

"W.P. Bidelman spoke of the need for a general reference catalogue giving full bibliographic data for individual stars. About 10^6 stars would be involved, and the resources of a large institution such as NASA would certainly be required to provide such a centralized information service. Several members of the Commission expressed interest in, and support for, this proposal."

(IAU Trans 1968, p. 220)

The NSF conference had its proceedings published in a special issue of the *Astronomical Journal* (Vol. **72/5**, June 1967, pp. 551-630). Although these contain a number of papers dealing with nomenclature, massive cataloguing, critical analysis, bibliographic issues, distribution, machine-readable compatibility, and so on, the recorded discussions fail to explicitly refer to the needs for data center facilities and activities, but these are definitely embedded "between the lines" and one could safely assume they were actually debated in informal talks between the attendees.

As a result of Perek's (1969) document, the IAU EC decided to set up an *ad hoc* working group (WG) and the following announcement was published in the IAU *Information Bulletin* dated June 1970, *i.e.* just before the 14th IAU GA in Brighton:

Working Group on Numerical Data in Astronomy and Astrophysics

The Chairman of the Working Group, Professor Ch. Moore-Sitterly, wishes to make the following announcement:

"The Executive Committee of the IAU has formed a Working Group to consider the problems of collecting and disseminating numerical data in astronomy and astrophysics. The membership of this Group is as follows:

Mrs. Charlotte Moore-Sitterly, Chairman
Prof. C.W. Allen[6]
Prof. W. Fricke
Prof. M. Migeotte
Prof. E. Schatzman

The Group must propose concrete steps to be taken in order to meet present and future demands for reliable numerical data in many specialized fields of astronomy and astrophysics.

This task calls for the active co-operation of the Presidents of all Commissions to handle or require collection of numerical data. It is essential to have concise reports from specialists regarding:

[6]Acting as Secretary at the meetings during the 14th IAU GA in Brighton.

– Existing information centers that are functioning well and operating on a continuing basis
– Information centers that are in the initial stage and should be encouraged to continue
– Specialized fields in which centers for numerical data should be established; and the data that are most utgently needed in each field.

Constructive suggestions for carrying out this mission on an Inter-Commission basis are solicited both before and during the Brighton Meeting."

(IAU IB 1970a, p. 14)

It should be precised that, at that stage, the working group was considered as temporary. This was going to change after the Brighton GA as its Resolution No. 7 stated that:

"Resolution No. 7

Proposed by the Working Group on Numerical Data
On the establishment of a Permanent Working Group on Numerical Data
The XIVth General Assembly of the International Astronomical Union resolves to establish a Permanent Working Group on Numerical Data."

(IAU IB 1970b, p. 5; IAU Trans 1971, p. 62)

More interestingly, a report of the WG meetings in Brighton was published (IAU Trans 1971, pp. 245-247). The opening general comments summarize well the discussions:

"General Comments

During the two sessions held by this Group, a lively and widespread interest in data handling was indicated. More than 80 persons attended each session and 16 countries were represented. The comments of 14 speakers, most of whom are operating data centers in various fields of research, stimulated active discussion. The serious need for coordinated effort and constructive planning with regard to Data Centers was striking. Ignorance of existing centers was apparent. Inter-Commission cooperation in publicizing all centers within the Union must be stimulated" (IAU Trans 1971, p. 245)

The following persons ("main data center leaders") were proposed as members of that first edition of the permanent WG on Numerical Data: G.A. Wilkins (Chairman), H.A. Abt, W.P. Bidelman, D.S. Evans, R.H. Garstang, R. Giacconi, B. Hauck, J.A. Hynek, C. Jaschek, J. Jung, T. Lederle, V.B. Nikonov, N.G. Roman, F.G. Smith, J.B. Sykes, H.M. Van Horn, and R. Wilson.

Historically very interesting, a snapshot of the data centers reported to the WG on 1 September 1970 was also provided as an appendix to the report with contact names and addresses:

"Data Centers Reported to Working Group 1 September 1970

– Atomic Energy Levels, W.C. Martin, Natl. Bur. Std., Washington, DC 20234, USA

– Cross Sections for Collisions of Electrons and Photons with Atoms, Ions and Small Molecules, L.J. Kieffer, Joint Institute for Laboratory Astrophysics, Boulder, CO 80302, USA

– Eclipsing Binaries, Photoelec. Obs. publ., continuing file of data, F.B. Wood, Univ. of Florida, Gainesville, FL 32601, USA

– Ephemerides, Planetary Data, Star Catalogues, G.A. Wilkins, Royal Greenwich Obs., Herstmonceux Castle, Hailsham, Sussex, UK – R.L. Duncombe, US Naval Obs., Washington, DC 20390, USA – Paris center

– Extragalactic Objects, J.D. Wray, Dearborn Obs., Evanston, IL 60201, USA

– Globular Clusters, Helen Sawyer-Hogg, David Dunlap Obs., Richmond Hill, Ontario, Canada

– Observatories, Instruments (Computer Listings) etc, A.G. Velghe, Obs. Royal de Belgique, Uccle-Brussels, Belgium

– Planetary Research Center, W.A. Baum, Lowell Obs., Flagstaff, AZ 86001, USA

– Pulsars-Center needed, Consult R. Giacconi, 11 Carleton St., Cambridge, MA 02142, USA

Star Catalogues, J. Jung, Stellar Data Center – Strasbourg
Radial Velocity Center – Marseille
T. Lederle, Astron. Rechen-Institut, Heidelberg, FRG
C.O.R. Jaschek, La Plata, Argentina
P.W. Hill, University Obs., St Andrews, Fife, UK
See, also, Ephemerides (above).

– Radio Sources, W.N. Brouw, Sterrewacht Leiden, Leiden, The Netherlands
Master List – Ohio State Univ.
Radio Astronomy – California Institute of Technology
Data on stars and galaxies

– Spectroscopic Binaries, A.H. Batten, Dominion Astrophys. Obs., Victoria, BC, Canada

– Transition Probabilities, W.L. Wiese, Natl. Bur. Std., Washington, DC 20234, USA" (IAU Trans 1971, p. 247)

Several comments are in order here:

– this is the first appearance, in an IAU document dated 1970, of the Strasbourg Data Center that would officially exist only in 1972, and next to the name of Jung who was going to be its first Director;

– several of the centers mentioned are not object-oriented and therefore will have destinies more specific to IAU Commissions than to the developments mentioned hereafter;

– it is difficult to assess what was exactly the degree of advancement of the projects listed above; other ventures such as the *thinking* mentioned separately (IAU Trans 1971, bottom p. 245) of an ASP[7] group under the chairmanship of H.A. Abt on the " feasibility and desirability of establishing an Astronomical Data Center" was not very deep and never went anywhere (Abt 2005);
– the multiplicity of (hyper)specialized "data centers" would persist in spite of the creation of much larger and structured pan-astronomical organizations; the proceedings of the conferences *Astronomy from Large Databases* (Murtagh & Heck 1988, Heck & Murtagh 1992) record the presentation of quite a number of such specialized data collections in existence a couple of decades later.

A triennium after the Brighton IAU GA, WG Chairman Wilkins reported on the group (IAU Trans 1973, pp. 757-760), starting with a statement agreed with the IAU EC on its activities:

> "The activities and membership of the Working Group shall be determined by the Executive Committee of the Union.
> The principal aims of the Group shall be:
> (i) to collect and publish information about the existence and preparation of files of numerical data, especially those in machine-readable form, that are relevant to the interests of the Union; and
> (ii) to recommend ways by which the preparation of such data files and the retrieval of information from them may be carried out economically and effectively, and to offer advice on these matters to appropriate Commissions.
> In carrying out these activities, the Group shall endeavor to complement, and not to duplicate, the work of the Commissions of the Union in their specialised fields." (IAU Trans 1973, p. 757)

The group obviously was going to be the place to exchange information on astronomical data matters, including those on the centers themselves. As stated however in the reports of the IAU Commission 5 [Documentation] meetings at the 15th IAU GA (IAU Trans 1974, pp. 73-74), the WG would be quickly attached to that Commission and renamed "WG on the Documentation, Exchange and Retrieval of Numerical data". Over time, it would go back to its shorter and much more handy label "Numerical Data".

Today it is still attached to IAU Commission 5 where this author chaired it after G.A. Wilkins (1970-1979), B. Hauck (1979-1984) and G. Westerhout (1984-1990). The WG Chairpersons are also the IAU representatives to CODATA where exchanges of information with colleagues of other scientific disciplines can take place.

The above chronology is summarized in Table 1.

[7]Astronomical Society of the Pacific.

TABLE 1. IAU chronology (until 1974) regarding astronomical data center projects worldwide.

1966	NSF conference on *The Construction and Use of Star Catalogues* (Univ. Maryland, 3-5 Oct 1966) (*Astron. J.* **72** (1967) 551-630)
1967	Prague IAU XIIIth GA – Commission 45 Meeting (IAU Trans 1968, p. 220)
1968	Jaschek's (1968) paper on *Information Problems in Astrophysics*
1969	IAU 32 Doc 11a on the urgency to coordinate actions on astronomical data (Perek 1969)
1970	14th IAU General Assembly (Brighton, 18-27 August 1970)
	Report of the preliminary WG on *Numerical Data for Astronomers and Astrophysicists* (IAU Trans 1971, pp. 245-248)
	Creation of a *permanent* IAU Working Group on *Numerical Data* (IAU Trans 1971, p. 62)
1973	Report of the permament WG on *Astronomical Data* (IAU Trans 1973, pp. 757-760)
1974	Report of IAU Commission 5 Meeting (24 Aug 1973) (IAU Trans 1974, pp. 73-74)

2. The Strasbourg Data Center

What was going on in France, and more particularly in Strasbourg, during all that time?

A few documents, including notes put together by P. Lacroute (1980), Director of the Observatory for the period 1946-1976[8], allow to get a good view of the progress towards the creation of a "Centre de Données Stellaires (CDS)" [Stellar Data Center][9].

The need for an organization able to collect and to disseminate stellar information was every day increasingly more urgent as data were piling up. Lacroute (1980) mentions A. Blaauw, R. Cayrel and J. Delhaye among the European astronomers pushing, in the late 1960s, for such a structure. Enquiries coming from abroad were circulated; surveys were carried out to review the issues (D.S. Evans on radial velocities; Ch. Moore-Sitterly on the collection of data carried out in Canada and USA; ...); it was appearing that

[8]For a general chronology of the Observatory history, see *e.g.* Heck (2005) in this volume.

[9]Renamed in the mid-1980s as "Centre de Données astronomiques de Strasbourg" to take into account it was, from then on, also dealing with non-stellar data.

many astronomers were interested in collecting information, but essentially only particular data or on specific type of stars.

The oldest trace we could find on the establishment of a stellar data center in Strasbourg goes back to the minutes of a meeting held in Strasbourg on 8 January 1969 with only six attendees from Paris and Strasbourg (Lévy, Delhaye, Lacroute, Jung, Artzner, Ratier). They discussed identification problems and which stellar data could be collected in astrometry as well as in other domains, calling in each case for the expertise of the most qualified astronomer.

As the idea of a data center was getting more precise, another meeting was organized in Strasbourg on 21-22 October 1969 in order to examine how in practice a data center could be made operational in Strasbourg with the best economy of means and the best efficiency:

> "Ce centre devrait pouvoir diffuser sous forme mécanographiquement exploitable, à tous les astronomes qui en feraient la demande, des données stellaires plus ou moins élaborées, regroupées par étoile. Ce centre devrait en outre signaler aux observateurs les lacunes les plus importantes dans les données existantes." (Lacroute 1969b)

> [That center should be able to disseminate, to all astronomers who would request it, machine-readable stellar data, more or less complex and star-oriented. That center should also point out, to the observers, the most important gaps in available data.]

The venture was becoming definitely international as the fifteen invited astronomers were coming from astronomical institutions in Geneva, Heidelberg, Marseilles, Paris and Strasbourg. Lacroute (1980) mentions that, around those times, J. Jung, *Astronome-Adjoint* at Paris Observatory and interested by the data center, was progressively installing himself in Strasbourg to devote his time to setting up the center.

In a letter dated 6 April 1970, Lacroute informed the IAU of the CDS project in Strasbourg and announced Jung as a speaker at the upcoming IAU GA in Brighton (August 1970 – cf previous section). Foreign institutions were then openly part of the venture:

> "A Strasbourg, nous envisageons d'organiser un Centre de Données Stellaires avec la coopération, notamment, de l'Observatoire de Heidelberg pour les positions méridiennes, de l'Observatoire de Lausanne pour la photométrie et de l'Observatoire de Marseille pour les vitesses radiales. Nous comptons que la mise au point du projet sera très avancée au moment du Congrès de Brighton." (Lacroute 1970a)

> [In Strasbourg, we are considering to set up a Stellar Data Center in cooperation with, among others, Heidelberg Observatory for the meridian positions, Lausanne Observatory for the photometry and Marseilles Observatory for the radial velocities. We reckon that the development of the project will be quite advanced at the time of the Brighton Conference.]

Another meeting was organized in Strasbourg on 4 July 1970 in order to review the status of the CDS project, to prepare the meetings at the upcoming IAU GA and to define priorities for the next steps of the work. The attendees were coming from the same institutions as the ones mentioned earlier, with the addition of Lausanne (Lacroute 1970b). The minutes of the meeting (Jung & Lacroute 1970) have a couple of interesting organizational comments:

> "Une structure définitive du Centre de Données et des tâches qui lui seront confiées ne pourra être décidée qu'après les discussions qui auront lieu à Brighton. Le Centre coordonnera en effet les activités d'Observatoires français et étrangers qui acceptent de participer à des travaux du type envisagé. Dès maintenant, une collaboration est établie entre les Observatoires de Lausanne (6 personnes), Marseille (1), Paris (3), Strasbourg (5); Heidelberg ayant donné son accord de principe. D'autres collaborations seront sollicitées."
>
> (Jung & Lacroute 1970)

> [A definitive structure for the Data Center and for its entrusted tasks will be decided only after the discussions that will take place in Brighton. The Center will coordinate the activities of French and foreign Observatories that will accept to take part in the activities of the considered profile. As for now, a collaboration has been established with the Observatories of Lausanne (6 persons), Marseilles (1), Paris (3), Strasbourg (5); Heidelberg has agreed in principle. Other collaborations will be sought for.]

After the IAU GA in Brighton, J.C. Pecker – who was involved in IAU Commission 5, but who was also the President *ad interim* of the *Comité National Français d'Astronomie (CNFA)* [French National Committee for Astronomy] – addressed a letter dated 19 October 1970 to the Director of the French *Institut National d'Astronomie et de Géophysique (INAG)* [10] [National Institute of Astronomy and Geophysics]. Pecker recalls the Brighton Resolution no. 7 as well as the favorable conditions for the CDS project, emphasizing the need to support it adequately. As it is again an excellent summary of the situation then, this letter is reproduced and its body translated hereafter:

19 Octobre 1970

Monsieur le Directeur de l'I.N.A.G.
Observatoire de Meudon
92 Meudon

Monsieur le Directeur,

L'Union Astronomique Internationale qui groupe les astronomes du monde entier a tenu cette année sa 14ème Assemblée Générale à Brighton, du 18 au 27 août. Dans la séance de clôture, parmi les 13 résolutions adoptées, je crois devoir vous signaler la résolution n° 7 par laquelle:

[10] Later renamed as *Institut National des Sciences de l'Univers (INSU).*

"La 14ème Assemblée Générale de l'Union Astronomique Internationale décide d'établir un groupe de travail permanent sur le problème des données numériques."

Au cours du Congrès, les travaux d'un groupe d'étude ont montré clairement l'importance fondamentale qu'il y a, pour l'avancement des recherches en astronomie, à regrouper les données numériques dans les différents domaines de l'astronomie, à les comparer et à les rendre commodément utilisables par les procédés automatiques modernes. Cette tâche primordiale est du reste difficile.

En France, certains astronomes n'ont pas attendu la résolution de l'Union Astronomique Internationale pour orienter ainsi leur activité. Déjà un catalogue bibliographique est en cours d'achèvement à Paris. Par ailleurs, depuis un an, l'Observatoire de Strasbourg a pris l'initiative de grouper la collaboration d'un nombre d'observatoires français et étrangers pour établir, à Strasbourg, un Centre de Données stellaires en vue de l'étude de notre galaxie.

Le but important, mais bien défini, et la répartition du travail entre les collaborateurs les plus qualifiés font bien augurer de la réussite de cette entreprise commune entre Strasbourg, Heidelberg, Lausanne, Genève, Marseille et Paris.

Le groupe de Strasbourg est du reste largement représenté dans le groupe de travail permanent constitué par l'Union Astronomique Internationale.

J'attire respectueusement votre attention sur l'intérêt qu'il y aura à appuyer cette entreprise internationale où le rôle de la France est important, par l'affectation de postes ou de crédits au moment où le besoin se fera sentir.

Veuillez agréer, Monsieur le Directeur, l'expression de ma considération distinguée,

J.C. Pecker
Président (par intérim) du Comité
National Français d'Astronomie

[The International Astronomical Union, which gathers together the astronomers of the whole world, held this year its 14th General Assembly in Brighton, from 18 to 27 August. In the closing session, among the 13 general resolutions adopted, I believe I have to call your attention on Resolution No. 7 saying:

"The 14th General Assembly of the International Astronomical Union resolves to establish a Permanent Working Group on Numerical Data."

In the course of the conference, the investigations of a study group clearly showed the fundamental importance, for the advancement of research in astronomy, to gather together numerical data in the various fields of astronomy, to compare them and to make them easily usable by automatic modern processes. Such a prime task is moreover difficult.

In France, some astronomers did not wait for the resolution of the International Astronomical Union to initiate such activities. A bibliographic catalogue is being completed in Paris. On the other hand, since one year, Strasbourg Observatory took the lead in setting up a collaboration between some French and foreign astronomers to establish, in Strasbourg, a stellar Data Center in view of studying our galaxy.

The important goal, well defined, and the distribution of work among the most qualified collaborators augur well of the success of that undertaking common to Strasbourg, Heidelberg, Lausanne, Geneva, Marseilles, and Paris.

The Strasbourg Group is actually largely represented in the permanent working group organized by the International Astronomical Union.

I call respectfully your attention on the interest to support such an international endeavor where the rôle of France is important, be it by assigning positions or budget lines whenever this will be necessary.]

The next CDS meeting was held in Geneva on 27 November 1970 where the dozen of attendees reviewed as usual the degree of advancement of their work. Decisions were also made on an *Information Bulletin*: to be published in English with a circulation of 600 copies (soon after brought up to 700) directed at astronomical institutions as well as at interested individuals. The first issue (May 1971) presented the Center's activities, reported on the participating observatories (Marseilles, Lausanne-Geneva, Heidelberg, La Plata) and published a couple of papers listing catalogues recently published. A second issue appeared in December 1971.

Lacroute (1980) indicates that, from the end of 1970, CDS started *de facto* to operate under the directorship of J. Jung. As to the personnel, there was the active collaboration of the "participating observatories" already mentioned and, in Strasbourg, the contribution from some members of the Observatory personnel. Some funding was received as INAG subventions; some punching of catalogs was carried out on remaining fund from a local astrometric project, but most money was injected by Strasbourg Observatory. The work performed was described in the widely distributed *Information Bulletin* as well as in the proceedings of a few conferences.

Meanwhile negociations between the Directors of INAG (J. Delhaye, very much interested in the center) and of Strasbourg Observatory (Lacroute) were progressively shaping up the by-laws for the data center. It was to become an INAG structure ensuring the stability requested by its activity. It would be installed at Strasbourg Observatory, at least at the beginning. On 12 May 1971, the CDS project was presented at the *Conseil de Direction* of INAG where it was well received, with however the recommendation that the computations carried on an INAG computer in Paris-Meudon find a local solution such as the CNRS computer at Strasbourg-Cronenbourg. On 26 May 1971, INAG's Scientific Council in turn examined the project, but requested more information on the data, on the methodologies as well on the way the personnel would be hired and employed. It recommended also additional presentations at specialized groups, wished that an INAG representative be part of the CDS Council and recommended also a participation of the personnel to the same Council.

At the next CDS Meeting (Strasbourg, 25 November 1971), Jung (1971) informed the Council about all those administrative delays, expecting INAG to create officially the Center in December 1971 after a Memorandum of Understanding (MoU) be signed between INAG and Strasbourg Observatory. This MoU was only approved by Strasbourg Observatory in its Council Meeting on 10 December 1971. The by-laws regulating CDS were approved at the beginning of 1972 and issued as INAG Decision 1/72/D0. Their first article specifies the aims and goals:

> L'INAG crée un Centre de Données Stellaires (CDS). Ce centre a pour buts, en vue d'études sur la Galaxie:
> 1. de rassembler les données les plus importantes sur les étoiles: positions, mouvements propres, magnitudes, spectres, parallaxes, etc.;
> 2. d'améliorer les données existantes grâce à leur confrontation;
> 3. de suggérer aux observateurs les observations les plus utiles pour compléter efficacement les informations disponibles;
> 4. de diffuser aussi largement que possible les résultats de ces travaux à la communauté astronomique;
> 5. d'effectuer des recherches sur la Galaxie à partir des données rassemblées.

[INAG creates a Stellar Data Center (CDS). This center has the following goals, in view of studying the Galaxy:

1. to collect the most important stellar data: positions, proper motions, magnitudes, spectra, parallaxes, and so on;
2. to improve the existing data by comparing them;
3. to suggest to observers the most useful observations to efficiently complete the available information;
4. to disseminate as widely as possible the results of those works to the astronomical community;
5. to undertake investigations on the Galaxy from the gathered data.]

The by-laws were completed by INAG decision 2/72/DO nominating the first CDS Scientific Council made of six French and six foreign astronomers[11]: A. Bijaoui (Nice Obs.), A. Blaauw (European Southern Obs.), J. Boulon (Paris Obs.), G. Cayrel de Strobel (Meudon Obs.), J. Delhaye (Paris Obs.), Ch. Fehrenbach (Haute Provence Obs.), W. Fricke (Astron. Rechen-Inst. Heidelberg), B. Hauck (Lausanne Inst. Astron.), C. Jaschek (La Plata Obs.), J. Jung (Strasbourg Obs.), G. Larsson-Leander (Lund Obs.), and C.A. Murray (Royal Greenwich Obs.) for a term of three years starting on 1 May 1972.

Their mission is specified in Article 5 of the by-laws:

> "Le Conseil Scientifique:
> – propose à l'INAG le Directeur du CDS;
> – oriente les travaux du CDS;
> – examine les projets de budget du CDS;
> – examine, en vue de leur transmission à l'INAG ou à d'autres organismes, les demandes de moyens présentées par le CDS;
> – approuve les rapports annuels du CDS."

[11]See the list of the successive CDS Councils as an appendix to this volume (Bruneau & Heck 2005).

[The Scientific Council:
– proposes to INAG the CDS Director;
– orients the CDS activities;
– examine the CDS budget projects;
– examine, in view of transmitting them to INAG or to other organizations, the requests for means presented by the CDS;
– approves the CDS annual activity reports.]

And finally INAG Decision 3/72/D0 dated 31 May 1972 was installing J. Jung as the first CDS Director. CDS was then officially and totally in business. The inaugural Council meeting took place on 25-26 May 1972. The following excerpt of the meeting minutes is of interest here:

> "J. Delhaye ouvre la séance en remerciant les membres d'avoir répondu à sa convocation, puis rappelle l'origine du projet de création d'un centre de données stellaires. Vieux projet, dont la paternité revient en partie à lui-même, mais aussi à A. Blaauw et à R. Cayrel, son exécution a été retardée pour des raisons administratives. Dans un souci de décentralisation, Strasbourg a finalement été choisi comme siège de ce CDS.
> Raisons de ce choix:
> – l'activité de P. Lacroute qui collaborait déjà avec W. Dieckvoss de Hambourg,
> – compétences du personnel.
> L'INAG, qui ne peut avoir de laboratoire propre, est l'agence qui définit la politique scientifique. Le CDS est une partie de l'Observatoire, mais a une ouverture internationale: aussi 50% des membres du Conseil Scientifique sont-ils des membres étrangers. Pour éviter d'en faire un centre uniquement rhénan, l'INAG a tenu à la participation de MM. Murray (Grande-Bretagne), Larsson-Leander (Suède) et Jaschek (Argentine)." (Jung 1972)

[J. Delhaye opens the session and thanks the [Council] members for having come to this meeting. He then recalls the origin of the project to set up a stellar data center. An old project indeed – the fatherhood of which is partly due to him, but also to A. Blaauw and R. Cayrel – the materialization of which has been delayed for administrative reasons. In a decentralization move, Strasbourg has finally been chosen as the CDS location.
Reasons for that choice:
– P. Lacroute's activity as he already collaborated with W. Dieckvoss from Hambourg,
– the personnel's competence.
INAG cannot have its own laboratories and is the agency that defines the scientific policies. CDS is part of [Strasbourg] Observatory, but has an international opening: therefore 50% of the CDS Scientific Council members are foreigners. To avoid making of CDS a Rhenish-only center, INAG wished the membership of Messrs. Murray (Great Britain), Larsson-Leander (Sweden) and Jaschek (Argentina).]

By the time of the CDS official existence, two issues of the *Information Bulletin* (May and December 1971) had already been published, with a third one scheduled for July 1972. The *Bulletin* would be published at an average rate of two per year until it was discontinued after Issue 48 (March 1996).

The pre-CDS chronology is summarized in Table 2. See also Jung (2005).

3. CDS' Success Story

CDS' success story has already been commented elsewhere (Heck 2000, 2002). Over three decades now, CDS moved from a file holder to an impressive information hub. It is recognized world-wide for the excellence of its work and its products, something also acknowledged through the collaborations established with CDS by the other astronomical data centers created subsequently[12].

It is out of the scope of this paper devoted to the pre-CDS years to review in detail the history of CDS. Interested readers can find more on the subsequent evolution in the successive issues of the *CDS Information Bulletin*[13]. For a description of the current activities and services, please refer to the CDS web site[14]. CDS has been an evolving structure, taking advantage of new media and often pioneering their usage before they became popular world-wide.

From a historical stand, it is interesting to investigate why CDS did so well. Since I saw CDS taking shape in the office where I was working at Paris Observatory in 1970-71[15] and because of my subsequent position as a closely-bound scientific user and, for some time, as the 'international CDS salesman', here are a few axes from first-hand testimony.

First of all, it must be stressed that the context around the creation of CDS was very different of what we are experiencing today with all those Virtual Observatory projects. There was no big machinery around nor big funding available. The general feeling was largely indifference, if not perhaps here and there some hostility because the motivations were not clearly understood. Such resentments would come up again later on from time to time, qualifying for instance data centers as *astrogroceries*.

CDS basic idea was a genial one: creating a huge table of identification synonyms giving access to all individual data from the integrated catalogs,

[12]See, for instance, *CDS Information Bulletin* Issue **14** (January 1978) p. 119 for NASA's Astronomical Data Center, Issue **15** (July 1978) p. 93 for Japan's Astronomical Data Service at Kanazawa Institute of Technology, and p. 98 for the Astronomical Council of the USSR Academy of Science – just to mention the first ones chronologically.

[13]Available through the link `http://adsabs.harvard.edu/` of the *Astrophysics Data System (ADS)*, following `Browse Library` with `BICDS` as journal code.

[14]`http://cdsweb.u-strasbg.fr/CDS.html`

[15]J. Jung, then at Paris Observatory, was my PhD thesis supervisor.

TABLE 2. Chronology of steps leading to the creation of the Stellar Data Center (CDS), officially installed at Strasbourg Observatory in May 1972.

1969	Meeting on future CDS (Strasbourg, 8 Jan 1969) (Lacroute 1969a)
	IAU 32 Doc 11. on urgency to coordinate actions on astronomical data (Perek 1969)
	Meeting on future CDS (Strasbourg, 21-22 Oct 1969) (Lacroute 1969b)
1970	Letter from Lacroute to IAU announcing a report on CDS at the upcoming IAU GA in Brighton (Lacroute 1970a)
	Meeting on future CDS (Strasbourg, 4 Jul 1970) (Lacroute 1970b, Jung & Lacroute 1970)
	14th IAU General Assembly (Brighton, 18-27 Aug 1970)
	Letter from CNFA Chairman to INAG Director (Pecker 1970)
	CDS Meeting (Geneva, 27 Nov 1970) (Jung 1970)
1971	CDS Information Bulletin # 1 (Ed. Jung, May 1971)
	Meeting of INAG's *Conseil de Direction* (Paris, 12 May 1971) (Aubert & Charvin 1971)
	Meeting of INAG's Scientific Council (Paris, 26 May 1971) (Boulon 1971)
	CDS Meeting (Strasbourg, 25 Nov 1971) (Jung 1971)
	CDS Information Bulletin # 2 (Ed. Jung, December 1971)
	MoU project between INAG and Strasbourg Observatory on CDS approved by Observatory Council (10 Dec 1971)
1972	INAG Decision 1/72/D0 creating CDS
	INAG Decision 2/72/D0 nominating the first CDS Council
	Inaugural Council meeting (25-26 May 1972) (Jung 1972)
	INAG Decision 3/72/DO nominating J. Jung as first CDS Director (31 May 1972)

all of this being completed by an object-oriented bibliography linked to the database.

If CDS' success is certainly due to the clearsightedness of its founders and to the web of collaborations set up *ab initio*, additional explanations can easily be found: right decisions taken at the right times by its initial managers, consistency of policies followed and, last but not least, the small but dedicated staff on civil-servant positions. The home-made software was certainly also an excellent vector, once made portable (even if this conversion was a very demanding exercise).

However, if most of the goals set by the founders have been reached[16], at least one of them did not really come through: astrophysical research directly geared to the center. This is worth stressing since such an ambition was explicitly written in the statutes. And this can probably be blamed onto the networks who allowed astronomers to stay at their home institutions rather than coming to CDS.

It is also probably correct and fair to say that, during its first years of existence, CDS was better perceived (or more successful) abroad than nationally. Things changed with the Hipparcos satellite, strongly supported by the French space agency CNES and with heavy French participation (see for instance Kovalevsky 2005). This consolidated for good the space involvment initiated with IUE and opened the way to a series of active collaborations with space and other large experiments.

In his 1973 report for the IAU WG on Numerical Data, Wilkins, among other interesting points, stressed the following ones:

> "The main factor that appears to influence the success, or otherwise, of a data centre is not its size, nor its hardware facilities, nor its location, but rather whether the persons responsible for maintaining the data files take an active interest in the generation and analysis of the data. When such an interest exists, any inadequacies in accuracy or content of the data will become apparent, and a continual improvement in the database and in retrieval facilities may be expected. The scope of a data centre should not be extended beyond the interests of the persons who are available at the centre, and the persons require both astronomical and data handling experience and knowledge."
>
> (IAU Trans 1973, p. 759)

A matter for meditation.

Acknowledgments

I am grateful to Jean Delhaye (1921-2001) and Pierre Lacroute (1906-1993) for the information they had provided over years on the CDS genesis, as well as to Chantal Bruneau and François Ochsenbein for their help in precising a couple of recollections.

References

1. Abt, H.A. 2005, personal communication.
2. Aubert, G. & Charvin, P. 1971, Minutes of INAG's *Conseil de Direction* Meeting held in Paris on 12 May 1971.

[16]And sometimes largely facilitated by not-so-quickly-expected technologies such as the electronic networking of the planet.

3. Boulon, J. 1971, Minutes of INAG's Scientific Council Meeting held in Paris on 26 May 1971.
4. Bruneau, Ch. & Heck, A. 2005, CDS Council Members, this volume.
5. Duerbeck, H.W. 2005, Walter F. Wislicenus and Modern Astronomical Bibliography, this volume.
6. Heck, A. 2000, From Data Files to Information Hubs: Beyond Technologies and Methodologies, in *Information Handling in Astronomy*, Ed. A. Heck, Kluwer Acad. Publ., Dordrecht (2000) 223-242.
7. Heck, A. 2002, The Impact of New Media on 20th-Century Astronomy, *Astron. Nahr.* **323** (2002) 542-547.
8. Heck, A. 2005, Strasbourg Astronomical Observatory and its Multinational History, this volume.
9. Heck, A. & Murtagh, F. (Eds.) 1992, Astronomy from Large Databases II, ESO Conf. & Workshop Proc. **43**, x + 534 pp. (ISBN 3-923524-47-1)
10. IAU IB 1970a, *Internat. Astron. Union Inform. Bull.* **24**.
11. IAU IB 1970b, *Internat. Astron. Union Inform. Bull.* **25**.
12. IAU Trans 1968, Transactions of the International Astronomical Union XIIIB, Proceedings of the Thirteenth General Assembly Prague 1967, Ed. L. Perek, D. Reidel Publ. Co., Dordrecht.
13. IAU Trans 1971, Transactions of the International Astronomical Union XIVB, Proceedings of the Fourteenth General Assembly Brighton 1970, Eds. C. de Jager & A. Jappel, D. Reidel Publ. Co., Dordrecht.
14. IAU Trans 1973, Transactions of the International Astronomical Union XVA, Reports on Astronomy, Ed. C. de Jager, D. Reidel Publ. Co., Dordrecht.
15. IAU Trans 1974, Transactions of the International Astronomical Union XVB, Proceedings of the Fifteenth General Assembly Poland 1973, Eds. G. Contopoulos & A. Jappel, D. Reidel Publ, Co., Dordrecht.
16. Jaschek, C. 1968, Information Problems in Astrophysics, *Publ. Astron. Soc. Pacific* **80**, 654.
17. Jung, J. 1970, Minutes of the CDS Meeting held in Geneva on 27 November 1970.
18. Jung, J. 1971, Minutes of the CDS Meeting held in Strasbourg on 25 November 1971.
19. Jung, J. 1972, Minutes of the CDS Meeting held in Strasbourg on 25-26 May 1972.
20. Jung, J. 2005, CDS: Origins and Early Beginnings, this volume.
21. Jung, J. & Lacroute, P. 1970, Minutes of the CDS Meeting held in Strasbourg on 4 July 1970.
22. Kovalevsky, J. 2005, The Hipparcos Project at Strasbourg Observatory, this volume.
23. Lacroute, P. 1969a, Minutes of the CDS Meeting held in Strasbourg on 8 January 1969.
24. Lacroute, P. 1969b, Agenda of the CDS Meeting to be held in Strasbourg on 21-22 October 1969.
25. Lacroute, P. 1970a, Letter to IAU (6 April 1970).
26. Lacroute, P. 1970b, Agenda of the CDS Meeting to be held in Strasbourg on 4 July 1970.
27. Lacroute, P. 1980, personal communication.
28. Murtagh, F. & Heck, A. (Eds.) 1988, Astronomy from Large Databases. Scientific Objectives and Methodological Approaches, ESO Conf. & Workshop Proc. **28**, xiv + 512 pp. (ISBN 3-923524-28-5)
29. Pecker, J.C. 1970, Letter to the Director of INAG (19 Oct 1970).
30. Perek, L. 1969, IAU EC 32 Doc. 11a (June 1969).

THE STELLAR DATA CENTER:
ORIGINS AND EARLY BEGINNINGS (1972-1974)

J. JUNG
22, rue Briant
F-92260 Fontenay-aux-Roses, France
jean.jung@thalesgroup.com

Abstract. In this chapter, the first Director of the Stellar Data Center (CDS) has put together a few recollections.

1. The CDS Idea

Astronomers have always catalogued stars. In the course of the long history of the discipline, it often happened that data collected without specific usage at the time were subsequently sources of essential discoveries in fields earlier unknown.

Such a tradition favoring long-term utility and – so one would say – the contemplation of stars since direct experimentation is excluded, enabled building up astrometric catalogs, progressively extended to other collections of observable quantities as new instrumental technologies became available (spectrography, photometry, and so on).

Difficulties however soon appeared as to the identification of the observed objects. Many astronomers broke new grounds in terms of designation systems, making the inter-connection of catalogs sometimes uselessly difficult (including at the time of computers).

One of the objectives upon the creation of the Stellar Data Center (CDS) was to improve as much as possible such a situation in order to valorize at best the existing data. Another aim stemmed from the fact that specialization of research and productivity requirements were ultimately leaving little availability to scientists for collecting data. For instance NASA, in spite of all its means, had, in the early 1970s, financed only one catalog of positions, the *Smithsonian Astrophysical Observatory Star Catalog* (Haramundanis 1966) that had however the enormous advantage to be then avail-

A. Heck (ed.), The Multinational History of Strasbourg Astronomical Observatory, 211–214.

able on magnetic tapes. The cost of digital coding (card punching then) of the various existing catalogs and of their (semi-automatic!) comparison was too high for people obliged to publish "scientific" results.

I had personally experienced such a situation: my PhD work – carried out at Paris Observatory at the end of the 1960s under the supervision of Jean Delhaye – was dealing with the statistical analysis of Dorrit Hoffleit's *Bright Stars Catalog* (1964), first compilation of astrometric and photometric data of some 10,000 stars. The cross-comparison with Gliese's catalog (1969) of nearby stars and others amply demonstrated the difficulties just mentioned.

The idea to create a center gathering astronomical catalogs and putting these onto machine-readable media usable by the scientific community at large came to light at about that time. It was directly in line with the French astronomical tradition that encompassed large projects such as the *Carte du Ciel*. It could also take advantage of the existence of a body of astronomers who, as somehow protected civil servants (or so to say secular monks ...), could devote their activities to long-term endeavors.

Credit for inspiring and setting up CDS goes to Jean Delhaye who managed to secure adquate funding in France and international support, including for Eastern Europe (Potsdam Observatory, for instance) and the Far East (Purple Mountain Observatory). One of the main contributors and supporters of the project was Geneva Observatory (under the directorship of Marcel Golay) and the associated Institute of Astronomy from the University of Lausanne (under the leadership of Bernard Hauck) dealing essentially with photometric data.

2. The Choice of Strasbourg

Once CDS's creation was accepted, a location had to be selected. In my opinion, Strasbourg Observatory has been retained for various reasons. First, the city is about equidistant to major centers such as Geneva, Heidelberg, Paris, Liège, ... It carries also a "supra-national" image resulting from the French-German alternance and more recently from the European Union. Strasbourg Observatory was also linked to the "Université Louis Pasteur" renowned in a number of disciplines (*e.g.* chemistry). Finally, the Observatory Director of the time, Pierre Lacroute, had novel approaches to astrometry that would be materialized in the Hipparcos space experiment[1]. Finally there were prospects of positions and promotions within the scope of a national policy of regionalization at the time.

When I took up duties in Strasbourg, the objective was clear: to establish a kind of *Rosette stone* for the existing catalogs, an unprejudiced

[1]See Kovalevsky's chapter in this volume. (Ed.)

reference that would simply be an interconnection of the main identification systems already in use and associated to observational data.

The real problem was rather to decide how to start: *a priori* it was sensible to start from a catalog rich enough to include more specialized ones. The AGK2 was available at the observatory, but digitizing it "manually" seemed to be an exaggerated task. OCR[2] technologies, then studied at CNRS[3] were disappointing, with an redhibitory error rate. The answer came from the USA where the SAO catalog, compiled for spacecraft navigation purpose, had just been made available on magnetic tapes. With its some 260,000 stars, it was ideally fuflilling our needs.

The work got quickly organized (with *i.a.* Marguerite Bischoff and Roger Bonnet) and the various astrometric catalogs were systematically connected to the SAO catalog, the data being punched on cards and then transferred onto magnetic tapes – all this leading ultimately to the constitution of one of the CDS's key products, the *Catalog of Stellar Identifications (CSI)*.

At the same time, additional staff joined the Observatory: collaborators improving the computer tool – I think mainly to Marc Wenger and François Ochsenbein who had imagine from the start an online distribution of the CDS data; or people exploiting the available data such as André Heck and Daniel Egret who extensively contributed to the renown of CDS, among others.

The "multidisciplinarity" of CDS was definitively achieved in 1974 with the arrival of Carlos Jaschek whose international notoriety helped to firmly establish the center as a resource accessible to all for reliable astronomical data.

3. Factors of Success

In the hindsight, it is easy to identify factors for success:

- resorting to a well-established structure ensuring continuity, *i.e.* state-sponsored research, independent from fashions and with guaranteed finances, even on a modest scale;
- undertaking nothing from the start without the actual support of the international community: the help of foreign observatories and NASA has been decisive;
- pragmatism: "manual" processing of tasks leading to effective results has always been favored compared to more "scientific" methods that might have sunk a start-up project;
- the support of the French national community and especially from Jean Delhaye.

[2]OCR = optical character recognition.
[3]CNRS = French National Center for Scientific Research.

References

1. Gliese, W. 1969, Catalogue of Nearby Stars, *Ver. Astron. Rechen Instituts Heidelberg* **22**.
2. Haramundanis, K.L. 1966, Smithsonian Astrophysical Observatory Star Catalog, Smithsonian Institution, Washington, DC.
3. Hoffleit, D. 1964, Yale Catalogue of Bright Stars, Yale Univ. Obs.

THE HIPPARCOS PROJECT AT STRASBOURG OBSERVATORY

J. KOVALEVSKY
Observatoire de la Côte d'Azur
Avenue Nicolas Copernic
F-06130 Grasse, France
jean.kovalevsky@obs-azur.fr

Abstract. The Hipparcos astrometric space mission is born at Strasbourg Observatory through the thoughts of its Director, Pierre Lacroute. This chapter describes the project prehistory and subsequent evolution, as well as the key rôle played by Lacroute.

1. The Prehistory of Space Astrometry

The prehistory of the Hipparcos astrometric space mission took place essentially at Strasbourg Observatory and more precisely through the thoughts on astrometry of its Director, Pierre Lacroute.

There was a striking need to improve the precision with which stellar positions, proper motions and parallaxes were determined in the 1960s. Only the nearest stars were showing position variations that could be significantly interpreted in terms of their space distribution and kinematics. Atmospheric deterioration limited stellar images to one arcsecond at best. Because of the measurement uncertainties on spots – often irregular – of that size, hopes for precision improvements were almost not significant (some 0.15 to 0.20 $''$).

Around 1965, Lacroute had started pondering that problem and had concluded the only way out was to point stars from artificial satellites. This was a daring idea for the time as space astronomy was then still in infancy. Requirements for mechanical, dynamical and thermal stability were still technologically out of reach. Nevertheless it was the right moment to start shaping the corresponding principles.

Another limitation of ground-based astrometry is the impossibility to observe the whole sky with one single instrument, while ultimately all stellar

A. Heck (ed.), The Multinational History of Strasbourg Astronomical Observatory, 215–219.

positions have to be referred to a unique frame covering the whole celestial sphere. For the first time, an instrument onboard an artificial satellite would allow to achieve that. But it was necessary to measure large angles.

The solution – a genial one since it is now applied for all new projects of astrometric satellites – was to use angle template made out of a complex mirror allowing to point two stellar fields with a constant reference angle. A grid located at the telescope focal plane would modulate the photon fluxes while the satellite would spin. The analysis of the modulations generated by two stars would allow to measure their angular distance.

2. The Project Progress

The first comprehensive document I know on this project dates back to 1966. It was a 33-page typed proposal, already very detailed, to the French space agency, the Centre National de Recherches Spatiales (CNES), for the study and possibly the materialization of the system Lacroute had conceived.

The following year, in August 1967, he presented his project to Commission 8 (Position Astronomy) of the International Astronomical Union (IAU) (Perek 1968, p. 63). He also submitted a new, more elaborated version to CNES including, among other new features, the suggestion to fly a prototype with a stratospheric balloon. This was however quickly rejected.

A feasibility study has indeed been carried out at CNES in 1969-70 (Husson 1974). At that time, the aim was to establish a position network made of at least 700 stars with a precision of the order of a $0.01''$. The study concluded that such a satellite, weighting 140kg and on a circular orbit at an altitude of more than 500km, could carry out the mission, but not with the desired precision.

The only way to improve things was to use a stabilized platform, something incompatible with the *Diamant* launcher. In any case, the French space program was stopped in 1971 and the only possibility was then to address the European Space Research Organization (ESRO). Presented in 1973 by the author, that project was selected together with twelve other proposals for a preliminary study.

In November 1973, Lacroute sent in a new project description, but, before studying it, ESRO organized a symposium (Frascati, October 1974) in order to assess the interest from the astronomy community in such a mission. The outcome being recognized as positive, a feasibility study was undertaken by the European Space Agency (ESA) that had succeeded ESRO.

3. The Data Reduction Challenge

During that time, Lacroute with the collaboration of Pierre Bacchus (in Strasbourg, then in Lille after 1973) continually improved the project and tackled also the challenge of the data reduction (Bacchus & Lacroute 1974). Bacchus analyzed the data reduction problematics and discussed particularly a method based on angular distances between stars (Bacchus 1975b), most appropriate for the configuration adopted Lacroute with a reference angle of 90° and a sky scanning by cercles intersecting each other at very small angles or for a mission restricted in time.

In addition, Lacroute was quite rightly worried about the instrument stabilization. He therefore proposed hosting the instrument onboard a TD1-type spacecraft, *i.e.* a polar satellite stabilized by gravity gradient. He studied also the possibility to launch a prototype on Spacelab, in spite of the short duration of the mission. A description of these two new versions was presented at the Frascati symposium (Lacroute 1975). For his part, Bacchus had proposed the optical configurations: an aplanetic system for the TD1 option and a Ritchey-Chrétien combination for the Spacelab one (Bacchus 1975a).

4. The Evolution of Lacroute's Ideas

It is interesting to illustrate the evolution of Lacroute's ideas – for instance about the structure of the complex mirror – as his thoughts were getting deeper and as the CNES study was progressing.

Fig. 1 shows the diversity of shapes considered between 1965 and 1974 for a reference angle always set at 90°. Note that the complex mirror ultimately adopted for Hipparcos is made of a circular mirror, with Schmidt corrections, cut in two pieces stuck together again to form a reference angle of 57°.

5. The Final Steps

ESA set up a working group gathering together astronomers (inluding Lacroute) and engineers from the agency for a definition study that concluded in May 1976. Then a new working group piloted a Study Phase A, step in which Lacroute continued to play an important rôle. With the same basic principles, but through more advanced technologies, the project was now heading for 25 times the number of stars aimed at by the 1973 proposal, and this with a precision 25 times better.

When the project was finally adopted by ESA under the name *Hipparcos*, Lacroute had become a member of the FAST Consortium Directing

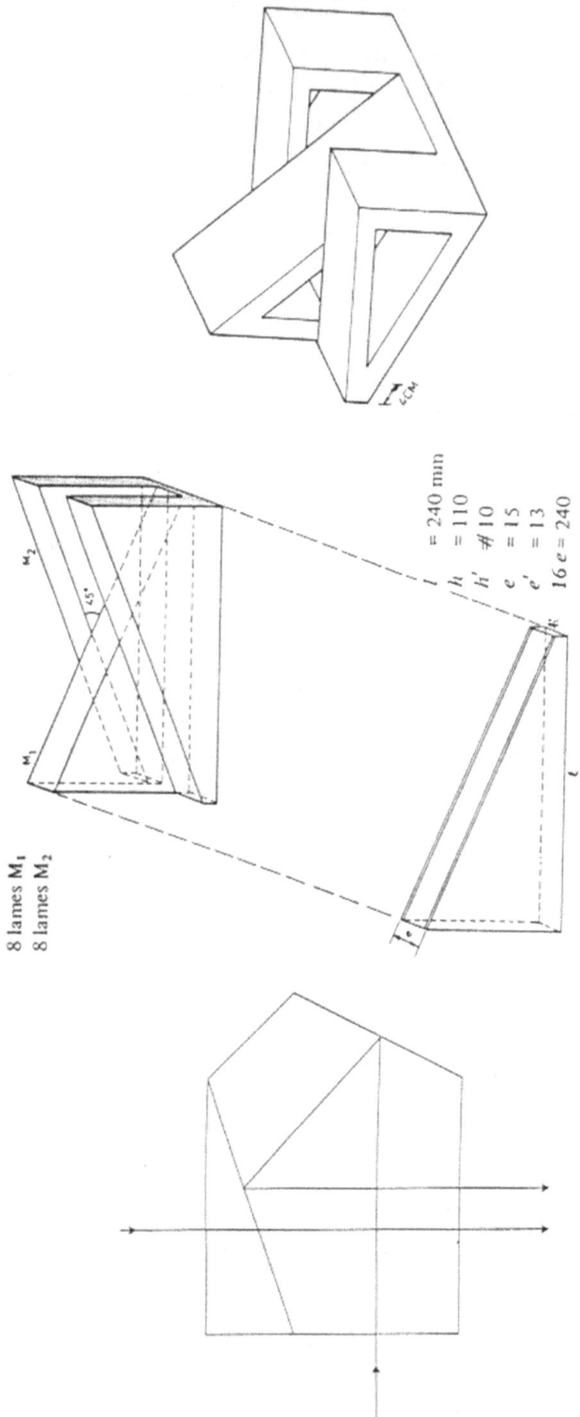

Figure 1. Succesive designs of the 90° reference angle for Hipparcos by Pierre Lacroute: left, a first sketch from 1966 (personal document); center, projet studied by CNES in 1970 (Husson 1975); right: project submitted to ESA in 1973 (Lacroute 1975).

Committee for the reduction of data including, under the author's leadership, teams from France, Germany, Italy, and the Netherlands.

Lacroute continued to take part actively in the reflections on the data reduction methodologies. In particular, he proposed to replace individual observations by functions expressing them over time intervals of several minutes (Lacroute 1983). Such a method, subsequently retained by the two data reduction consortia, improved very significantly the consistency of the results and decreased the uncertainties.

Lacroute's last enjoyment has been witnessing the launch of "his" satellite and being aware of its success. Here is a sentence from his thanking address after being nominated as *Personnalité de l'Année* (Person of the Year) in April 1991:

> *"Sometimes people call me "Hipparcos' father": this is partially true since I gave birth to it and I got devoted to it like a child; one must put some passion in one's undertakings, but one must remain lucid. In fact, after some time, the child education has been entrusted to younger and more competent persons. This is quite normal and one can only be pleased with that. Hipparcos is really a very collective work."*

References

1. Bacchus, P. 1975a, Space Astrometry Project: Optical Concept, in *Space Astrometry*, Eds. T.D. Nguyen & B.T. Battrick, European Space Research Org. **SP-108**, 31-35.
2. Bacchus, P. 1975b, Space Astrometry Project: Reduction by Multilateration, in *Space Astrometry*, Eds. T.D. Nguyen & B.T. Battrick, European Space Research Org. **SP-108**, 37-42.
3. Bacchus, P. & Lacroute, P. 1974, Prospects of Space Astrometry, in *New Problems in Astrometry*, Eds. W. Gliese, C.A. Murray & R.H. Tucker, Reidel Publ. Co., Dordrecht, 277-282.
4. Husson, J.C. 1975, Feasibility of an Astrometrical Satellite – Results of a CNES Study, in *Space Astrometry*, Eds. T.D. Nguyen & B.T. Battrick, European Space Research Org. **SP-108**, 43-47.
5. Lacroute, P. 1975, Space Astrometry Projects, in *Space Astrometry*, Eds. T.D. Nguyen & B.T. Battrick, European Space Research Org. **SP-108**, 5-30.
6. Lacroute, P. 1983, Le Lissage d'Attitude dans le Project Hipparcos, in *Processing of Scientific Data from the ESA Astrometry Satellite Hipparcos – Thinkshop of the FAST Consortium*, Ed. P.L. Bernacca, Univ. Padova, p. 141.
7. Lacroute, P. 1991, Nécessité de l'Astrométrie Spatiale, *Mém. Acad. Sc. Arts Belles-Lettres Dijon* **132**, 97-100.
8. Perek, L. (Ed.) 1968, Commission de l'Astronomie de Position, in *Proceedings of the Thirteenth General Assembly, Prague 1967* **XIIIB**, Reidel. Publ. Co., Dordrecht, 61-64.

STRASBOURG OBSERVATORY AND THE ASTRONOMISCHE GESELLSCHAFT

W. SEGGEWISS
Observatorium Hoher List
Universitäts-Sternwarte Bonn
D-54550 Daun/Eifel, Germany
seggewis@astro.uni-bonn.de

Abstract. In September 1881 and a century later, in February 1982, the *Astronomische Gesellschaft (AG)* met at Strasbourg Observatory. The 1881 meeting, the 9th AG General Assembly, focussed essentially on the international *Zonen-Unternehmen* while the 1982 Spring Meeting gathered together astronomers from Austria, Germany, France and Switzerland around discussions on the Milky Way structure and kinematics.

1. The Astronomische Gesellschaft

The *Astronomische Gesellschaft* (Astronomical Society, hereafter AG) was founded in 1863 in Heidelberg by 26 young astronomers. Though mostly of German nationality, the founders intended to establish a truly international society. Indeed, up to the First World War (WWI) only about one third of the members were German.

The most important scientific work that the AG initiated soon after its foundation was the so-called *Zonen-Unternehmen* which resulted in the famous zone catalogues AGK1 (**A***stronomische* **G***esellschaft* **K***atalog* No. 1, 1890-1924), AGK2 (1947-1958), and AGK3 (1975). Only seven German observatories appear among the twenty institutes at which observations were conducted. This reflects the high international reputation of the society.

2. The 9th General Assembly (1881) in Strasbourg

After the Franco-German War of 1870-71, science and teaching were rapidly re-established in Strasbourg: in 1872 the university was founded and in 1875 the construction of a new observatory began. The main instrument,

A. Heck (ed.), The Multinational History of Strasbourg Astronomical Observatory, 221–225.

the 50cm Repsold refractor, saw first light in 1880. In order to mark the official inauguration of the Observatory, the Director, A.F.Th. Winnecke, former Vice-Director of Pulkovo Observatory, invited the AG to hold its 9th General Assembly from 22 to 24 September 1881 in Strasbourg.

Fifty-four members of the society, all respectable men[1], met for their business and science sessions in the grand hall of the refractor building (Big Dome). The number of participants – 54 – was small compared to the total number of 304 society members. But in those days travelling was a bit more tedious and expensive than today and meetings were reserved mainly for the "big" astronomers, *e.g.* "Messieurs les Directeurs" H.G. Bakhuyzen (Leiden), W. Foerster (Berlin), H. Gyldén (Stockholm), P.J.-C. Janssen (Meudon), F. Küstner (Bonn), S. Newcomb (Washington), J.A.C. Oudemans (Utrecht), G.V. Schiaparelli (Milano), O. Struve (Pulkovo), and E. Weiss (Vienna).

A. Auwers was AG President and, *ex officio*, Chairman of all sessions. The main point of discussion was, of course, the progress of the *Zonen-Unternehmen.* Eleven observatories had sent progress reports, among them E.C. Pickering for the zone 55° – 50° of Cambridge, MA. We learn that most observations of the program stars were finished and that they were now diligently measuring the stars of the *Fundamental Catalog (FC)* which A. Auwers (1879) had just compiled.

Another topic would keep the astronomers busy for the following century: the standardization of stellar magnitudes. Photographic photometry could not have been meant, because it was only ten years earlier that the British physician Maddox had invented the dry bromine-silver-gelatine emulsion. A committee of the *American Association for the Advancement of Arts and Sciences* submitted the proposal to divide the stars in three classes:

I. Naked-eye stars.
II. Bright telescopic stars.
III. Faint telescopic stars.

The first class was assigned to the *Royal Astronomical Society*, the second one to the AG, and the third one to the American Association, *i.e.* to North-American observatories, to establish 24 sequences of standard stars along the celestial equator. In addition, two sequences near the North pole were to be observed. The assembly took no decision, but the President promised to continue the negotiations with the North Americans.

Two other more technical topics were discussed. First, the assembly decided that, in the *Astronomische Nachrichten* and in all AG publications, comets should finally be designated with year plus Roman number in the

[1]It was not until 40 years later, in 1921, that the first female astronomers were admitted into the AG.

Figure 1. Arthur Auwers (1838-1915), President of the *Astronomische Gesellschaft* at the time of 1881 General Assembly of the society in Strasbourg. Auwers also authored the first release of the *Fundamental Catalog* (1979). (© Astron. Gesellschaft)

sequence of perihelion passage in that year. This is still largely used in astronomy! Second, the assembly agreed that the Smithsonian Institution should hosts the *Central Bureau of Astronomical Telegrams* – as it still does today. But in those early days the cipher-code looked fairly complicated as one can read in Appendix X of the conference proceedings (Schönfeld & Winnecke 1881, p. 344).

The members in the grand hall listened to about 30 lectures including presentations of celestial globes, small instruments and ground plans of new observatories. In this respect, two Belgian colleagues showed drawings of the projected Royal Observatory at Brussels and of the Observatory at Liège which was still under construction. Judging from today, the most important lecture was given by J. Janssen (Paris) under the title *"Photographie de la Comète 1881 III"*. Janssen (1881) enthusiastically advertised the use of photographic photometry in astronomy. Indeed, it was the time when an observer like Janssen did not hesitate to write:

> *"J'ai eu l'honneur de mettre sous les yeux de la société internationale d'astronomie une photographie de la Comète 1881 III".*

3. The 1982 Spring Meeting

The great wars of the 20th century brought big disruptions to the AG history. Up to the WWI, the society, albeit with its seat in Germany, was recognized as the only truly international community of astronomers. But after the war, in 1919, the *International Astronomical Union (IAU)* was founded "to promote and safeguard the science of astronomy in all its aspects through international cooperation"" (see the IAU homepage[2]). The AG still had many non-German members, *e.g.* E. Strömgren (Copenhagen) was President from 1921 to 1930. J.C. Kapteyn (Groningen) and A.S. Eddington (Cambridge) acted as Vice-Presidents for several years. But important international matters were now dealt with by the new IAU.

The Second World War (WWII) brought a complete breakdown of the AG (see *e.g.* Pfau 2000/OSA 1). It had to be refounded in 1947 when A. Unsöld (Kiel) became President. Nowadays, the members originate mainly from the German-speaking countries in middle Europe, *i.e.* from Austria, Germany, and, partly, from Switzerland.

In 1982, A. Florsch, then Director of Strasbourg Observatory, and C. Jaschek, Director of the *Centre de Données Stellaires*, invited the AG to hold a Spring meeting in Strasbourg which had become a European capital – celebrating the 100th anniversary of the observatory and about 100 years after the first AG meeting in the city. Invitations also were sent to the *Société Française des Spécialistes d'Astronomie* and to the *Schweizerische Gesellschaft für Astrophysik und Astronomie.* 130 astronomers met at Strasbourg, among which 21 colleagues from France. G.A. Tammann (Basel) was then AG President and Chairman of the meeting. The general topic of the meeting was the *"Structure and Kinematics of the Galaxy"*.

The outstanding event was the Karl-Schwarzschild-Vortrag (Lecture) delivered by J. Delhaye, past Director of Paris Observatory and of the French *Institut National d'Astronomie et de Géophysique (INAG).* In 1959, the AG had established the Karl-Schwarzschild-Preis in honour of this eminent astronomer and his scientific achievements. J. Delhaye's lecture summarized *"Die Bewegungen der Sterne und ihre Bedeutung für die galaktische Astronomie"*. A well-attended public lecture was held by Ch. Fehrenbach from Haute Provence Observatory on *"Les Grosses Planètes"*. Five surveys, given by colleagues invited from France, Germany, and Switzerland, provided an introduction to the topic of the meeting. They were supplemented by 40 contributed papers and 4 (four!) posters. Indeed, the presentation of scientific results on posters was just in the starting phase.

It is noteworthy that the hosts at Strasbourg had organized a varied and interesting program of social events – in strong contrast to the spartan

[2]`http://www.iau.org/`

meeting of 1881, which fully concentrated on science alone. The author wishes to mention particularly an organ concert at the Cathedral and an overwhelming closing dinner with all the typical Alsatian dishes and plenty of Edelzwicker wine ...

A Note on the Literature

The proceedings of the 1881 General Assembly have been published by E. Schönfeld (Bonn) and A. Winnecke (Strasbourg), at that time the two (!) AG Secretaries. The proceedings of the 1982 meeting have been edited by the author (Seggewiss 1982a), Secretary from 1980 to 1986. The history of the *Astronomische Gesellschaft* has been written by F. Schmeidler (1988). A special essay on all AG General Assemblies up to 1981 has also been produced by the author (Seggewiss 1982b). The AG is also maintaining a web site[3].

References

1. Auwers, A. 1879, Fundamental-Catalog für die Zonen-Beobachtungen am nördlichen Himmel, *Publ. Astron. Ges.* **14**, 48 pp.
2. Janssen, J. 1881, Photographie de la Comète 1881 III., *Vierteljahrsschr. Astron. Ges.* **16**, 308-311.
3. Pfau, W. 2000, The Astronomische Gesellschaft: Pieces from its History, in *Organizations and Strategies in Astronomy*, Ed. A. Heck, Kluwer Acad. Publ., Dordrecht, 67-75.
4. Schmeidler, F. 1988, Die Geschichte der Astronomischen Gesellschaft. Jubiläumsband – 125 Jahre Astronomische Gesellschaft, Astron. Ges. Hamburg, vi + 70 pp.
5. Schönfeld, E. & Winnecke, A. (Eds) 1881, *Vierteljahrsschrift Astron. Ges.* **16/4**.
6. Seggewiss, W. (Ed.) 1982a, *Mitt. Astron. Ges.* **57**, 119-331.
7. Seggewiss, W. 1982b, Die Versammlungen der Astronomischen Gesellschaft 1863-1981, *Mitt. Astron. Ges.* **57**, 135-142.

[3] `http://www.astro.uni-jena.de/Astron_Ges/ag0home.html`

STRASBOURG OBSERVATORY AND ASTRONOMICAL EVENTS IN THE REGIONAL PRESS IN THE EARLY 20TH CENTURY

W. BODENMULLER
Dernières Nouvelles d'Alsace
17-21 rue de la Nuée Bleue
F-67077 Strasbourg Cedex, France
w.bodenmuller@dna.fr

AND

A. HECK
Observatoire Astronomique
11, rue de l'Université
F-67000 Strasbourg, France
heck@astro.u-strasbg.fr

Abstract. This chapter illustrates rare appearances in the regional press of Strasbourg Observatory around the beginning of the 20th century. The particular cases discussed are the Observatory inauguration in 1881, a fireball triggering some 150 letters addressed to the Observatory in 1904, and bright comets around 1910.

1. Introduction

The discovery in Strasbourg Observatory archives of a voluminous file dealing with a bright meteor in the Alsatian sky on 21 March 1904 led us to look for mentions of the event in the local press and more particularly in the *Straßburger Neueste Nachrichten (SNN)* [Strasbourg Latest News] that became subsequently in French the *Dernières Nouvelles d'Alsace (DNA)* [Latest News from Alsace][1].

[1]Alsace is the smallest region of France made of two *Départements* (Bas-Rhin and Haut-Rhin) hosting a population of about 1,750,000 inhabitants over 8280 square kilometers (Year 2000 numbers), but the *Dernières Nouvelles d'Alsace (DNA)* is one of the most important regional daily newspapers in France (more than 200,000 copies) with 26 local editions including one largely in German.

A. Heck (ed.), The Multinational History of Strasbourg Astronomical Observatory, 227–244.

Starting from that celestial phenomenon that struck people's minds, we attempted to find traces of other astronomical events such as the appearances of bright comets at the beginning of the 20th century or, a bit earlier, the inauguration of the Imperial Observatory in 1881.

2. The 1904 Meteor File

The stack of old yellowed letters is really impressive. Most of them were in German Gothic handwriting [Kurrentschrift] and addressed to the then Director of the Imperial Observatory Ernest Becker (1843-1912) and/or to his Assistant Karl Schiller (1882-1979). The file, held for a century in the Observatory archives, survived the conflicts that rocked the area subsequently. Its 150 pieces are dealing with a bright meteor, very likely a fireball, over the Rhine valley on 21 March 1904.

What triggered such a massive correspondence? Was the reputation of the place so big, while it had been inaugurated only some twenty years earlier? The considerable quantity of documents was certainly the result, at least partially, of a call published by the Observatory in the SNN on 24 March (Fig. 1), three days after the event that took place on 21 March around 20:30. The Observatory invited witnesses of the phenomenon to communicate details such as the exact time, the direction, the course in the sky, the appearance, and so on. The same call appeared also the same day in another daily newspaper of the city, the *Straßburger Post* (Fig. 2). We suspect, from the contents of the letters, that it was also echoed in other regional newspapers, possibly via press agencies.

A first mention of the fireball was however printed already on 22 March in the SNN: a few lines speaking of a *"a big blue ball, leaving behind a wavy trail."* No report of the phenomenon appeared in the issue dated 23 March, but, the following day, two small notes related observations made at Pisdorf and at Zeinheim, two villages near Saverne, some 40km Northwest of Strasbourg. The meteor was again mentioned a last time on 28 March and then disappeared from the newspaper.

Those few, rather brief, references in the press are in contrast with the numerous letters sent to the Observatory between 24 March and 4 April 1904. Two features from these are striking. First, precise observations with numerous details were frequently accompanied by sketches or carefully drawn illustrations. About a century ago (no television!), people were probably paying more attention to celestial phenomena which were better watched than today. The light pollution over our cities nowadays and the everyday flood of images are certainly playing some rôle in this.

Samples of letters are illustrated in Figs. 3 to 6. They show the variety of input (mostly in German, but also a few in French) from shop owners,

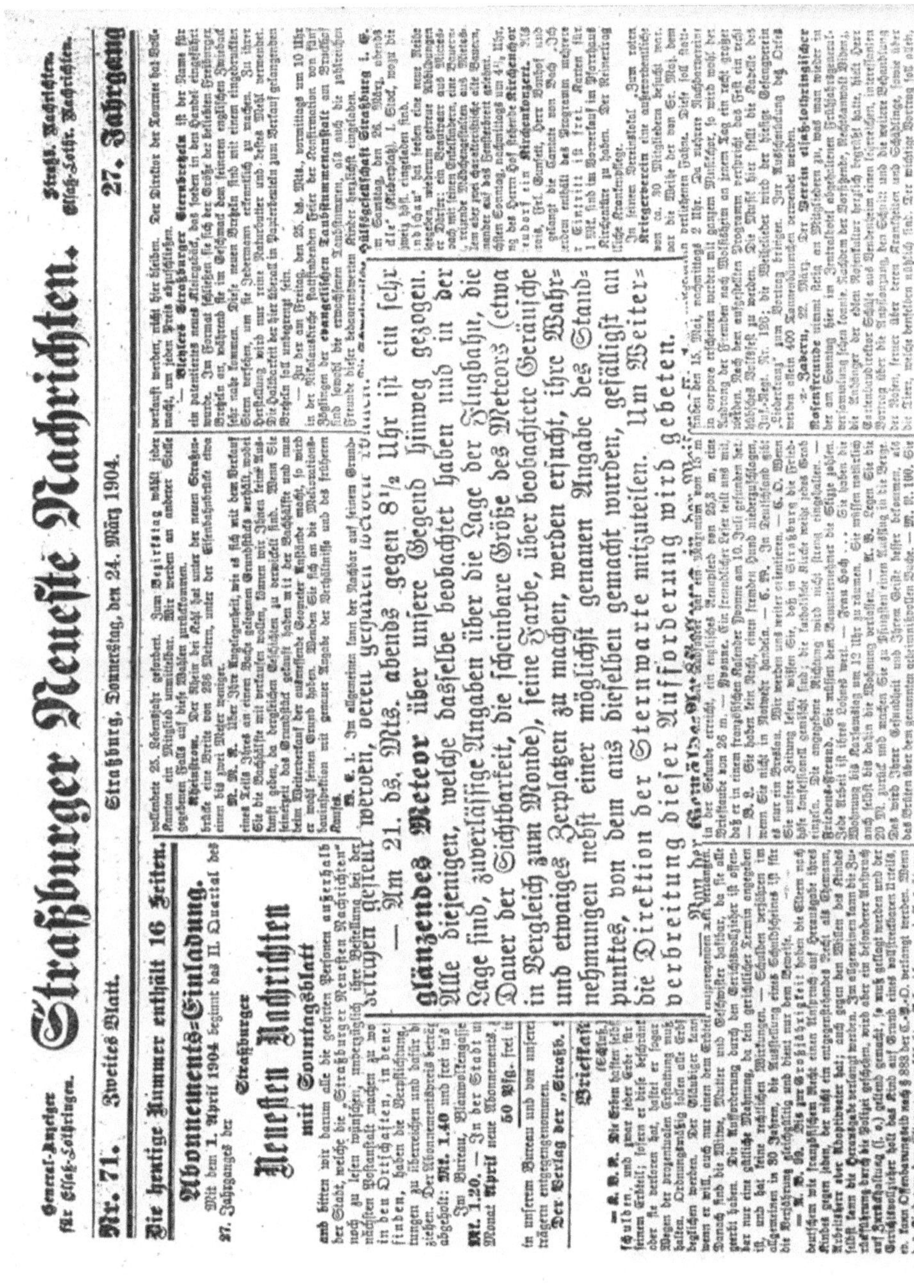

Nr. 71. Zweites Blatt.

Straßburger Neueste Nachrichten.

Straßburg, Donnerstag, den 24. März 1904.

27. Jahrgang

— Um 21. ds. Mts. abends gegen 8½ Uhr ist ein sehr glänzendes Meteor über unsere Gegend hinweg gezogen. Alle diejenigen, welche dasselbe beobachtet haben und in der Lage sind, zuverlässige Angaben über die Lage der Flugbahn, die Dauer der Sichtbarkeit, die scheinbare Größe des Meteors (etwa in Vergleich zum Monde), seine Farbe, über beobachtete Geräusche und etwaiges Zerplatzen zu machen, werden ersucht, ihre Wahrnehmungen nebst einer möglichst genauen Angabe des Standpunktes, von dem aus dieselben gemacht wurden, gefälligst an die Direktion der Sternwarte mitzuteilen. Um Weiterverbreitung dieser Aufforderung wird gebeten.

Figure 1. The call published by Strasbourg Imperial Observatory in the *Straßburger Neueste Nachrichten (SNN)* on 24 March 1904 following the bright meteor of 21 March.

Donnerstag, 24. März. Zweite Morgen-Ausgabe. 1904. Nr. — 307.

Straßburger Post.

Vierteljährlicher Bezugspreis: in Deutschland M

Anzeigenpreis die Zeile oder deren R

Straßburger Stadtnachrichten.

△ Straßburg, 23. März.

Wie bereits mehrfach gemeldet, ist am 21. d. M. abends gegen 8½ Uhr ein sehr glänzendes Meteor über unsere Gegend hinweggezogen. Alle diejenigen, welche dasselbe beobachtet haben und in der Lage sind, zuverlässige Angaben über die Lage der Flugbahn, die Dauer der Sichtbarkeit, die scheinbare Größe des Meteors (etwa im Vergleich zum Monde), seine Farbe, über beobachtete Geräusche und etwaiges Zerplatzen zu machen, werden ersucht, ihre Wahrnehmungen nebst einer möglichst genauen Angabe des Standpunktes, von dem aus dieselben gemacht wurden, gefälligst an die Direktion der Sternwarte mitzuteilen. Um Weiterverbreitung dieser Aufforderung wird gebeten.

Wir lesen in der „Kölnischen Zeitung": Auch der „Fidelio" bewies

Figure 2. The call as published in another city newspaper, the *Straßburger Post* on 24 March 1904.

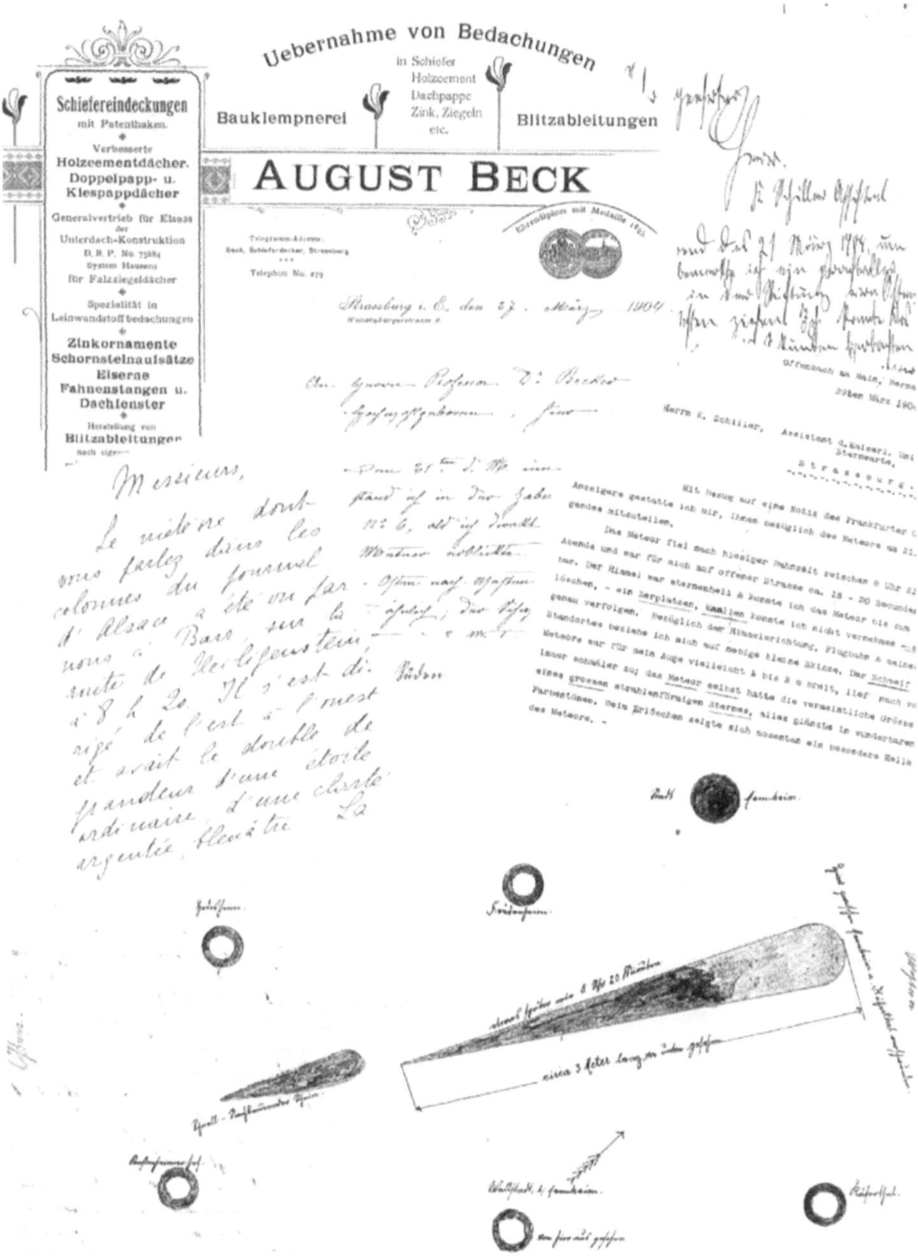

Figure 3. Examples of letters received at Strasbourg Observatory about the 1904 fireball. Note they address either Becker (Director) or Schiller (Assistant). The letter on the middle left is in French.

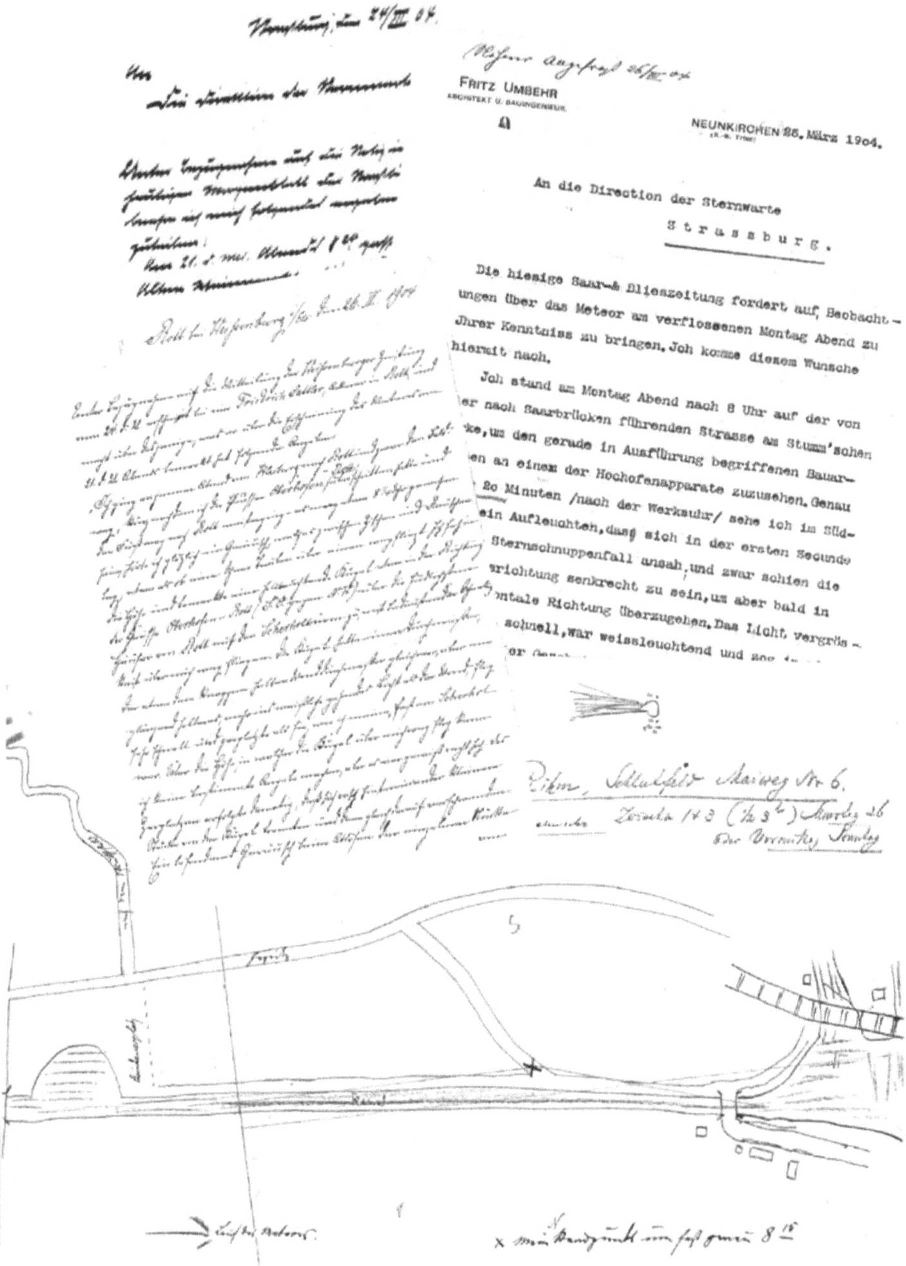
FRITZ UMBEHR
ARCHITEKT U. BAUINGENIEUR.

NEUNKIRCHEN 25. März 1904.

An die Direction der Sternwarte

S t r a s s b u r g .

Die hiesige Saar-& Blieszeitung fordert auf, Beobacht- ungen über das Meteor am verflossenen Montag Abend zu Jhrer Kenntniss zu bringen. Joh komme diesem Wunsche hiermit nach.

Joh stand am Montag Abend nach 8 Uhr auf der von ...er nach Saarbrücken führenden Strasse am Stumm'schen ...ke, um den gerade in Ausführung begriffenen Bauar... ...en an einem der Hochofenapparate zuzusehen. Genau ... 20 Minuten /nach der Werksuhr/ sehe ich im Süd- ... ein Aufleuchten, dass sich in der ersten Secunde ... Sternschnuppenfall ansah, und zwar schien die ...richtung senkrecht zu sein, um aber bald in ...ntale Richtung überzugehen. Das Licht vergrös- ... schnell, war weissleuchtend und ...

Figure 4. More examples of letters on the 1904 fireball. Nowadays only few people can decipher handwritten German Gothic such as in the top left letter. The typewriter used for the top right letter was missing the eszet (ß replaced by ss – see Footnote 2).

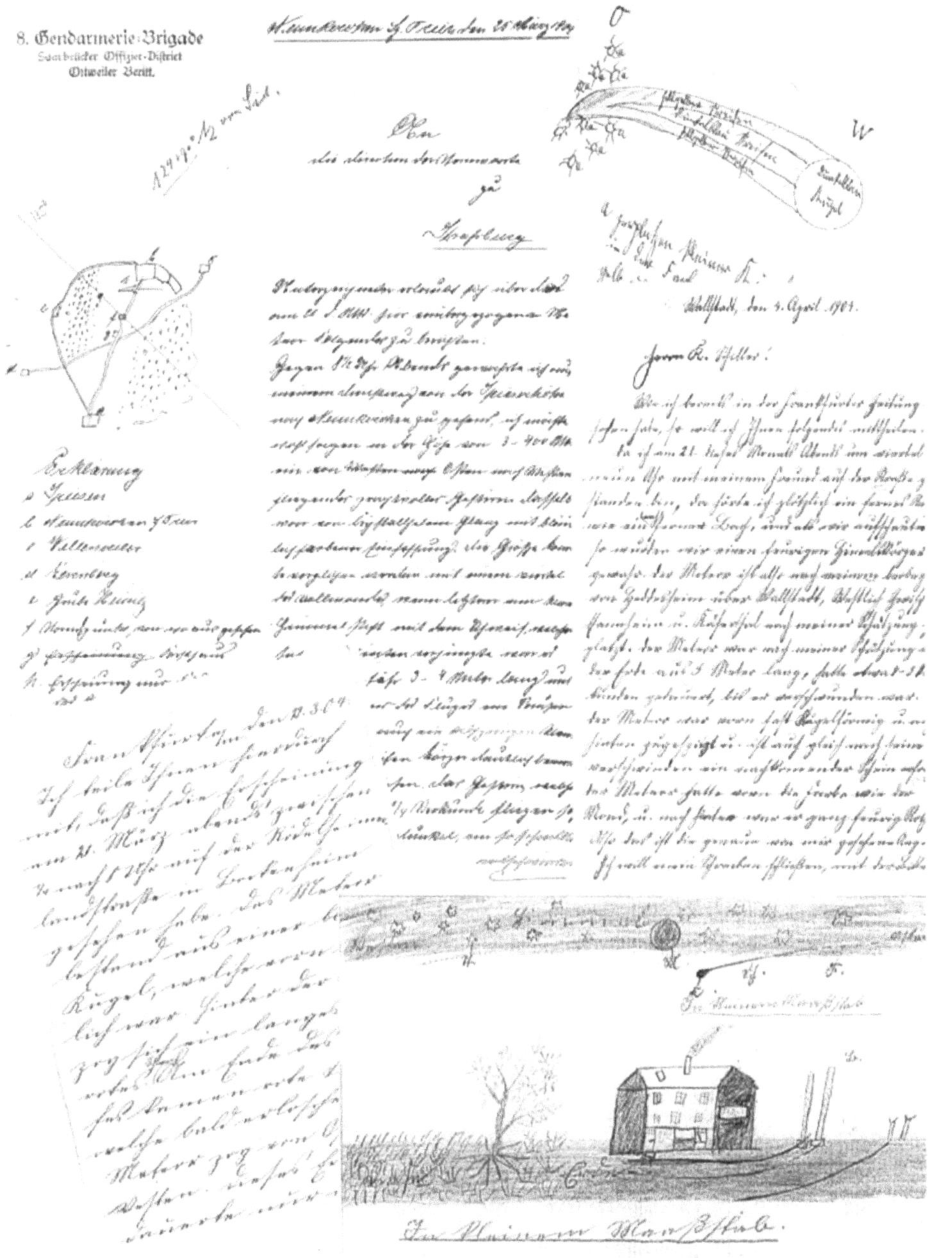

Figure 5. More examples of letters on the 1904 fireball. The top left letter, structured as a real police report, came indeed from a state police post.

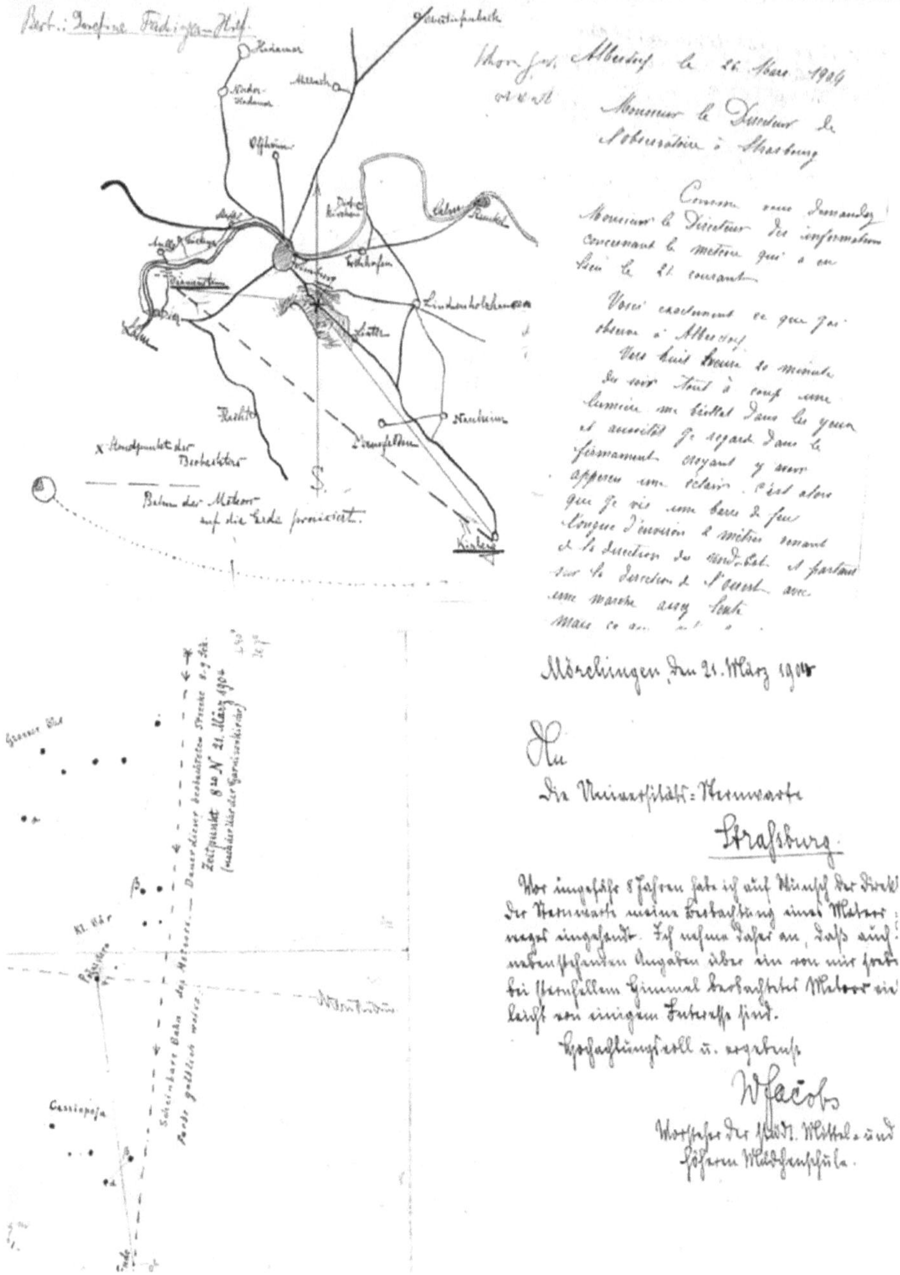

Figure 6. More examples of letters on the 1904 fireball. Admire the top left drawing. The top right letter is in French.

police officers, parish priests, commoners and peasants. Note the typed letters[2] at a time when etiquette required people to write letters with the neatest hand and liquid ink.

Fig. 7 reproduces over two pages an especially interesting communication sent by a locomotive stoker on a large sheet of drawing paper[3].

The second surprise came from the extent of the area from which originated the letters: from Strasbourg and the Alsatian countryside of course, but also from Lorraine, from the Palatinate and the Saar, from Luxembourg and Mainz, and so on. Why did those senders address their observations to Strasbourg rather than to Heidelberg or Berlin? The attraction of Strasbourg, the 'new', attractive and bursting city of the time was perhaps part of the explanation, as well as possible additional appearances of the Observatory call, echoed in other newspapers thanks to press agencies.

Apparently the Observatory never published in the newspapers a synthesis of all the mail received and we failed to find anything in the Observatory annual reports and in the *Annalen.* But an interesting series of short notes appeared in the *Astronomische Nahrichten.*

H. Rosenberg (1905a) first described a method for determining the trajectory of meteors and subsequently applied it to the 21 March 1904 one (Rosenberg 1905b). This triggered some comments from P. Moschick (1905) of Heidelberg Observatory, answered in turn by Rosenberg (1905c).

According to Rosenberg (1905b), the meteor lit up around Weil-der-Stadt (Kepler's birthplace) near Stuttgart at about 57km of altitude. The

[2]Typewriters were then in infancy. Remington started producing their first typewriter on 1 March 1873 in Ilion, NY. Another early typewriter manufacturer was Underwood. One could guess the typewriters used here had been imported from the US since the typical German character ß was missing and had to be typed as `ss`.

[3]The framed text appearing on the righ-hand side of Fig. 7 can be translated as follows:
"Right Honorable Sir,
Hereby I am sending you a sketch of the trajectory and of the observation of the meteor that I happened to be able to observe by chance. As I guessed that a call would appear in the newspaper, I took note precisely of everything. Perhaps you will be able to use my data.
From the train station clock, it was 8:25 [pm] when the meteor showed up, from *a*-*b* on the drawing, illuminating with full light, brighter than the moonshine. It was a white light mixed with blue rays. In *b*, it went out and splitted up until *e* into red-glowing pieces that scattered in all directions. I perceived also a slight hissing and a noise of crackling. The whole event lasted 22-25 seconds. My position of observation was the road to Mittelhausbergen in Cronenbourg.
With respectful greetings from
Franz Brockow
Qualified (Locomotive) Stoker."
This text was deciphered from German Gothic and translated into French by Louis Ludes who discovered in the process he was currently living in a house next to the one occupied then by the Brockow family!

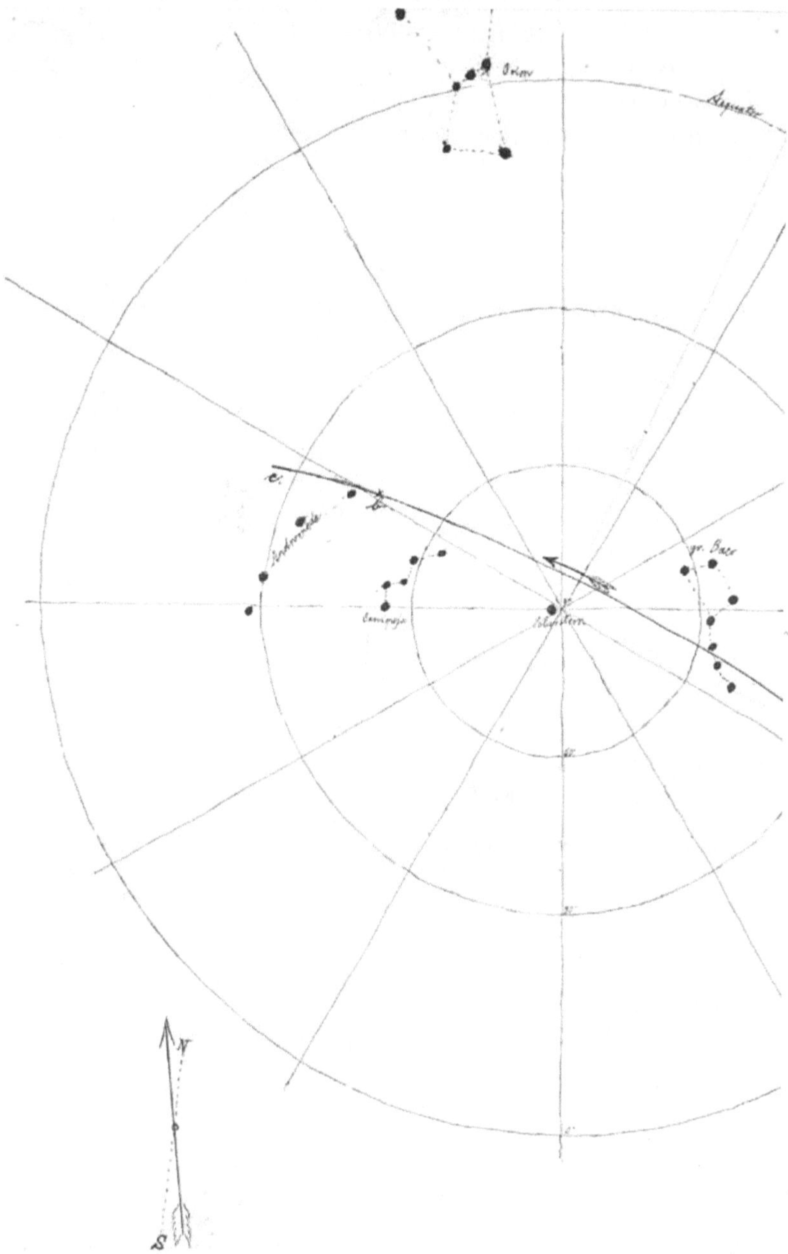

Figure 7. An especially interesting and detailed drawing received from a locomotive stoker called Franz Brockow living near Strasbourg. See Footnote 3 for a translation of the framed text appearing on the right-hand page as well as for an interesting anecdote.

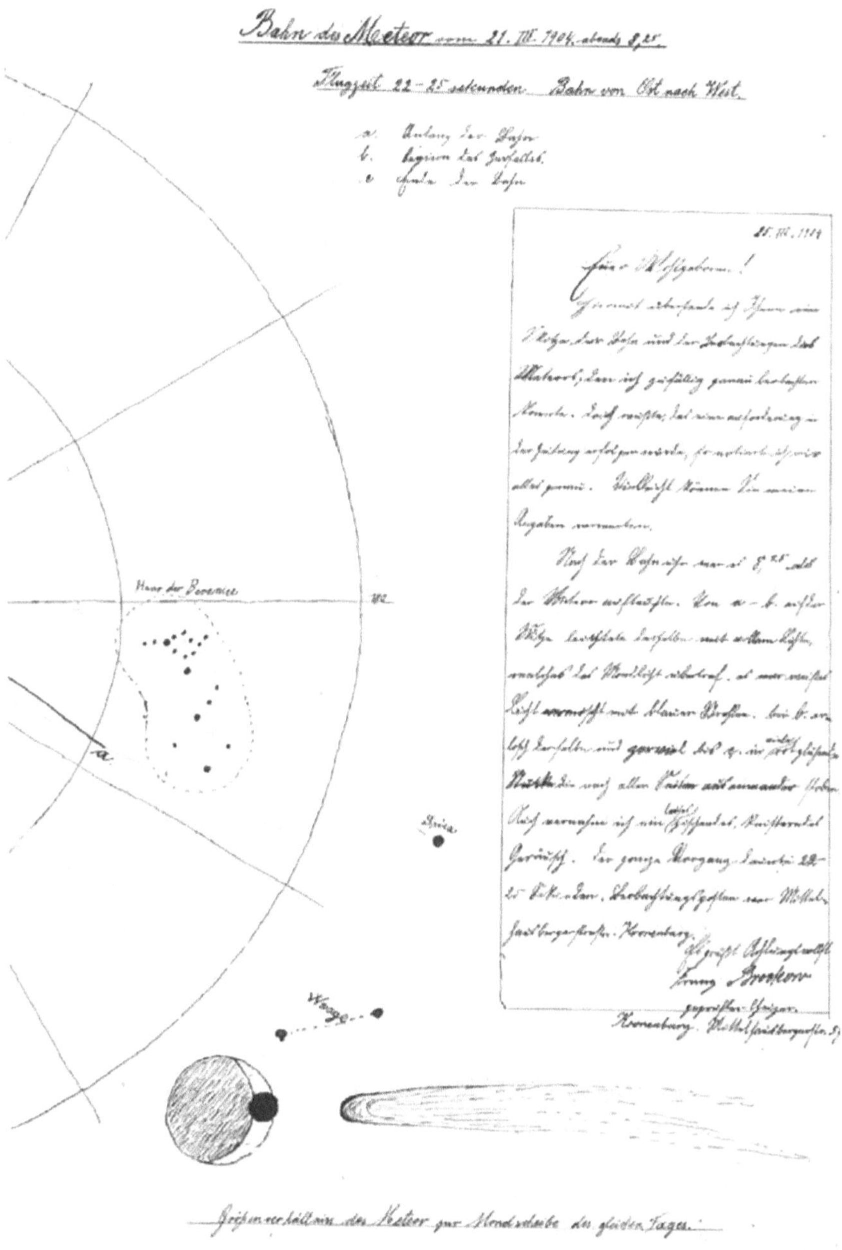
Bahn des Meteor vom 21. IV. 1904, abends 8,25
Flugzeit 22 – 25 sekunden Bahn von Ost nach West.
Haar der Berenice
Spica
Waage

object was moving roughly Westwards at an average speed of about 41km/h until splitting up some 20' North of Rheims (the crowning city of French Kings) at a final altitude of about 126km. This would mean that the meteorite was not falling towards the ground, but in fact going up, after apparently rebounding on the densest atmospheric layers in a ricochet effect.

Moschick offered different values obtained with a different method and argued about the influence of uncertainties. But he basically confirmed the ascending movement.

Amazingly enough, at the time of concluding this chapter, a fireball appeared in the Alsatian sky, following a ground trajectory similar to the 1904 one, but this time with a usual descending movement, triggering hundreds of testimonies – both spontaneous and answering calls from local associations and planetariums (Bodenmuller 2005). On 18 February 2005 towards 22:30, the meteorite was seen as coming from Germany, crossing Westwards the Alsatian plain at an altitude of about 50km, passing a few kilometers North of Vittel and likely hitting the ground around the border between the *départements* of Vosges and Haute-Marne where the impact is currently being searched.

3. Other Astronomy-Related Events

From the wealth of observations of the March 1904 fireball and the vast area covered by the letters, we wondered whether there would be more mentions, in the SNN, of the Observatory activities or of other astronomical events.

Since it would take ages to flip through all pages of the tens of volumes in the newspaper archives[4], we chose to tackle first a few specific slots such as the Observatory inauguration in September 1881 and the appearance of bright comets: Comet Morehouse (1908c/1908 III) in October/November 1908, the "Great Comet" (1910a/1910 I) of January 1910[5], Halley's Comet in May 1910 and Comet Brooks (1911c/1911 V) in October/November 1911.

3.1. THE OBSERVATORY INAUGURATION

The edification of Strasbourg Imperial Observatory was initiated in 1877, leading to an official inauguration by a General Assembly of the *Astronomische Gesellschaft (AG)* from 22 to 24 September 1881.

"It was then the largest and the most modern observatory in Europe", say the DNA on 17 May 1947 in a long article devoted to the difficulties of the Observatory after World War II. As recalled elsewhere in this volume

[4] All issues have been microfilmed since the newspaper foundation in 1877.

[5] Also called Johannesburg Comet as it had been discovered in South Africa.

Gesammt-Auflage: 12.533 Exemplare.

Neueste Nachrichten

DERNIÈRES NOUVELLES

Nr. 225. Straßburg, Sonntag, den 25. September 1881. 4. Jahrgang.

... die Geschichte und Physik erstrecken.

Straßburg, 23. Sept. In der neuen Sternwarte wurde gestern die 9. Jahresversammlung der Astronomischen Gesellschaft abgehalten und hatten sich zu derselben eine beträchtliche Anzahl Theilnehmer, auch von auswärts, eingefunden. Ihnen zu Ehren fand Abends im Stadttheater eine Festvorstellung statt und gelangte die Mozart'sche Oper „Figaro's Hochzeit" zur Aufführung.

Straßburg, 24. Sept. Wie der „M. Z." von ihrem ...

Figure 8. The seven lines reporting the General Assembly of the *Astronomische Gesellschaft* inaugurating Strasbourg Imperial Observatory in the *Straßburger Neueste Nachrichten (SNN)* on 25 September 1881. Note the mention of the socio-cultural event, attending the *Nozze di Figaro* [Figaro's Hochzeit] and the subheader in French. (© DNA)

(Seggewiss 2005), the AG held another General Assembly at the Observatory in 1982 to celebrate its centenary, an event that was also reported in the DNA (21 & 22 February 1982).

Well, in 1881, the inauguration of the largest and most modern European observatory went almost unnoticed, at least in the SNN. Seven modest lines on the first page of the 25 September issue (Fig. 8) informed that the yearly assembly of the *Astronomische Gesellschaft* was held at the new observatory *"with a large number of participants, many of which came from abroad"* and that these attended in the evening a performance: Mozart's *Nozze di Figaro.*

Meanwhile, the newspapers from 20 to 27 September 1881 reported at length on an agricultural exhibition in Strasbourg, on the Kaiser's visit in Karlsruhe where he was attending a princely wedding, as well as on the re-opening of the schools taking place that Monday 26 September 1881!

3.2. COMETS

Jumping a quarter of century ahead in time, to Fall 1908, four years and half after the 1904 fireball, we would expect the region was enjoying the celestial display of Comet Morehouse (1908c/1908 III). Camille Flammarion's *Astronomie Populaire* re-edited in 1955 by André Danjon[6] displays magnificent pictures of that comet taken in October 1908 at Meudon Observatory.

From 14 to 26 October 1908, there was no mention of that hairy object in the SNN, but we learn that five professors of Strasbourg University had been invited to attend Prince August Wilhelm's wedding in Berlin, that dramatic rescue operations took place at a race of montgolfiers, and that the Zeppelin airship had carried out two splendid flights.

Also the seismic station[7], located at a stone throw from the Observatory, is frequently mentioned as it recorded earthquakes in Turkmenistan, some 6,000 km away from Strasbourg, on 14 and 16 October 1908. A week later, another quake struck Plauen's area in Southwest Saxony.

The situation changed fifteen months later with the unexpected arrival of a new bright comet (Great January Comet 1910a/1910 I) while people were already preparing themselves for the return of Halley's Comet, predicted for April/May 1910. In their issue of 29 January 1910, the SNN devoted a long article to the history of comets, reminding that the last bright comet was visible in 1882. On 1 February 1910, the newspaper mentioned

[6]Danjon was then Director of Paris Observatory after having been in charge of Strasbourg Observatory from 1929 to 1945.

[7]See Duerbeck's (2005) contribution to this volume for a section on the beginnings of seismology in Strasbourg under the leadership of Ernst von Rebeur-Paschwitz (1861-1895).

Gelesenste, verbreitetste Zeitung Elsaß-Lothringens.

Straßburger Neueste Nachrichten

Haupt-Insertionsorgan des Reichslandes.

mit achtseitiger Gratis-Beilage: „Sonntagsblatt“.

Straßburger Nachrichten. — Elsaß-Lothringische Nachrichten.

General-Anzeiger für Straßburg und Elsaß-Lothringen.

Nr. 113. Erstes Blatt. Straßburg, Mittwoch den 18. Mai 1910. 33. Jahrgang.

Der Komet.

Im Zeichen der Kometen.

Figure 9. Top part of one of the articles on comets run by the *SNN* around Halley's Comet return in 1910, here in the issue of 18 May 1910. (© DNA)

observations carried out at the *Kehltor* [gate to Kehl, the city across the Rhine], at Pisdorf near Saverne[8], in Sarralbe, as well as at the Archenhold Observatory in Berlin-Treptow.

Strasbourg Observatory was mentioned nowhere, neither with comet observations, nor with explanations. A similar situation occurred around mid-May 1910 when the comet fever triggered by Halley's Comet return reached its climax: not a single day between 17 and 27 May passed without a paper on Halley's Comet or on the passage of the Earth through its tail, something anxiously expected; observations were reported from Heidelberg, Potsdam and Vienna; but no word was given from Strasbourg Observatory where the Director was now Julius Bauschinger (1860-1934), having taken over from Becker in 1909.

On 21 and 23 May 1910, an article's author regretted that the comet – perhaps hidden by the haze above the horizon – was still invisible from Strasbourg. Finally, on 24 May, it was visible, but the observer was rather disappointed and, on 27 May, the tail was reported as still not visible from Mulhouse nor Strasbourg, in spite of the usage of instruments. As to the astronomers from the Imperial Observatory, they seemed to be ... eclipsed too from the press. On the contrary, the seismic station was present again in the news with a remote earthquake in the Aleutian Islands and, on 27 May 1910, a much closer quake shaking the Sundgau (Southern Alsace).

After such a disappointing performance from Halley's Comet (actually much alike the 1985-86 return), what could be expected of the nevertheless beautiful Comet Brooks (1991c/1911 V) visiting us in October 1911? Between 28 October and 3 November 1911, the hairy object remained unreported in the newspaper that announced happily an aviation day in Strasbourg on 5 November with seven or eight montgolfiers ...

Strasbourg imperial astronomers were however not quite inactive as testified by the Observatory reports. Thus Director Becker (1909) mentioned 27 observations of Comet 1908c (Morehouse), probably carried out at the comet seeker illustrated in Fig. 10 reprinted from Krisch's (1901) *Lexikon*[9]. In the next report, the new Director Bauschinger (1910) mentioned modestly three observatorions of Comet 1909a, three more of 1909e and amazingly only four observations of Halley's Comet, but his following report (Bauschinger 1911) is much more generous with 35 observations of Halley's Comet, eight of Comet 1910a (Great Comet), ten of Comet 1910b and two of Comet 1910c. The subsequent annual reports are silent on comet observations.

[8]A century ago, a sky aficionado was definitely living in Pisdorf since observations of the 1904 fireball were already reported from that village (see above).

[9]See also Heck (2005) in this volume for a picture of the same instrument used in the 1920s by a French astronomer, possibly Rougier, on the Big Dome terrasse.

Figure 10. Strasbourg Observatory comet seeker as illustrated in Krisch's (1901) *Astronomisches Lexikon.* See also Footnote 9.

4. Last Comments

From the above examples, one cannot conclude that Strasbourg Observatory was totally absent from the media at the beginning of the 20th century, but its silence is somehow astonishing if compared to our current standards.

Acknowledgments

We gratefully acknowledge the assistance of DNA Head Archivist Joël Fréchard, of Observatory Librarian Philippe Vonflie, and of Louis Ludes for deciphering Franz Brockow's handwriting.

References

1. Bauschinger, J. 1910, Straßburg [Observatory Annual Report], *Vierteljahrsschr. Astron. Ges.* **45**, 263-267.
2. Bauschinger, J. 1911, Straßburg [Observatory Annual Report], *Vierteljahrsschr. Astron. Ges.* **46**, 189-192.
3. Becker, E. 1909, Straßburg [Observatory Annual Report], *Vierteljahrsschr. Astron. Ges.* **44**, 277-285.
4. Bodenmuller, W. 2005, Une "boule de feu" dans le ciel alsacien, *Dernières Nouvelles d'Alsace* (25 Feb 2005).
5. Duerbeck, H.W. 2005, The Observatory of the Emperor Wilhelm University: The People behind the Documents, this volume.
6. Flammarion, G.C. & Danjon, A. 1955, Astronomie Populaire Camille Flammarion, Flammarion, Paris, 674 pp.
7. Heck, A. 2005, Strasbourg Observatory and its Multinational History, this volume.
8. Krisch, A. 1901, Astronomisches Lexikon, Hartleben's Verlag, Wien/Pest/Leipzig, viii + 632 pp.
9. Moschick, P. 1905, Bermekungen zu dem Artikel "Heller Meteor von 1904 März 21" in Astr. Nahr. 4008, *Astron. Nahr.* **168**, 39-40.
10. Rosenberg, H. 1905a, Über eine Methode zur Bestimmung von Meteorbahnen, *Astron. Nahr.* **167**, 49-53.
11. Rosenberg, H. 1905b, Heller meteor von 1904 März 21, *Astron. Nahr.* **167**, 385-390.
12. Rosenberg, H. 1905c, Heller Meteor von 1904 März 21, *Astron. Nahr.* **169**, 105-108.
13. Seggewiss, W. 2005, Strasbourg Observatory and the Astronomische Gesellschaft, this volume.

THE CORONELLI GLOBE OF STRASBOURG OBSERVATORY

A. HECK
Observatoire Astronomique
11, rue de l'Université
F-67000 Strasbourg, France
heck@astro.u-strasbg.fr

Abstract. Strasbourg Observatory is hosting a damaged Coronelli 1693 celestial globe appearing in the 1920s for the first time in Observatory documents. This chapter reviews Coronelli's activities and elaborates on the history and possible trajectory of that globe.

1. Sharing Office with a Coronelli Globe

A Coronelli celestial globe is discernable on a photograph from the 1920s (Fig. 1, reproduced from Esclangon 1926) showing the rotunda of the Big Dome of Strasbourg Observatory. The globe was still in the same area in the early 1980s: a kind of toy that every passer-by felt obliged to put the hands on and to rotate. By then, the upper part was already blackened by dust, dirt, finger grease and whatever had fallen on the globe over the years, making it impossible to decypher some texts and drawings from the Northern hemisphere (compare for instance Figs. 2 & 3). A large portion of the lower part had been torn off and bits were missing here and there (Figs. 4 & 5).

In order to protect the globe from further degradation, it was moved into my (locked) office when I arrived in early 1983 at Strasbourg Observatory. That large room (then basically unused) was located in the left wing when entering the rotunda. After renovation, that room became the Director's office when I took up duties as such in 1987 (with then the Observatory Secretariat installed in the right wing, just across the rotunda in the old meeting room). At the time of writing these lines (early 2005), the globe is still in the Director's office and in the same conditions.

A. Heck (ed.), The Multinational History of Strasbourg Astronomical Observatory, 245–254.

Figure 1. The *Rotonde*, circular hall of the Big Dome, in the 1920s (Esclangon 1926). The 1693 celestial globe of Vincenzo Coronelli (1650-1718) is visible on the left. (© Obs. Astron. Strasbourg)

2. Coronelli's Globe Industry

Vincenzo Coronelli (Fig. 6) was born in Venice on 15 August 1650. At the age of 15, he entered the convent of the Minor Brothers of San Nicolò della Lattuca and, at 21, the great convent of Santa Maria Gloriosa dei Frari where he would established in 1685 an etching workshop very active in cartograhic production. Meanwhile he went first to Rome (Santo Bonaventura College) where he received his degree in theology.

His cartographic career really took off in 1678 when he went to Parma to build a couple of 175cm[1] globes for Duke Ranuccio Farnese. These drew the attention of César Cardinal d'Estrées[2] who invited Coronelli to Paris. Thus, in 1681, Coronelli arrived in the French capital with the task of building two 384cm globes – one terrestrial, one celestial – for King Louis XIV. This was commissioned by d'Estrées, apparently following a suggestion by Jacques Borelly (unknown birthdate – 1689) from the recently created Academy of

[1]The sizes given in this chapter correspond to the diameters of the globes.
[2]His brother, François-Annibal 1er d'Estrées, was French Ambassador in Rome.

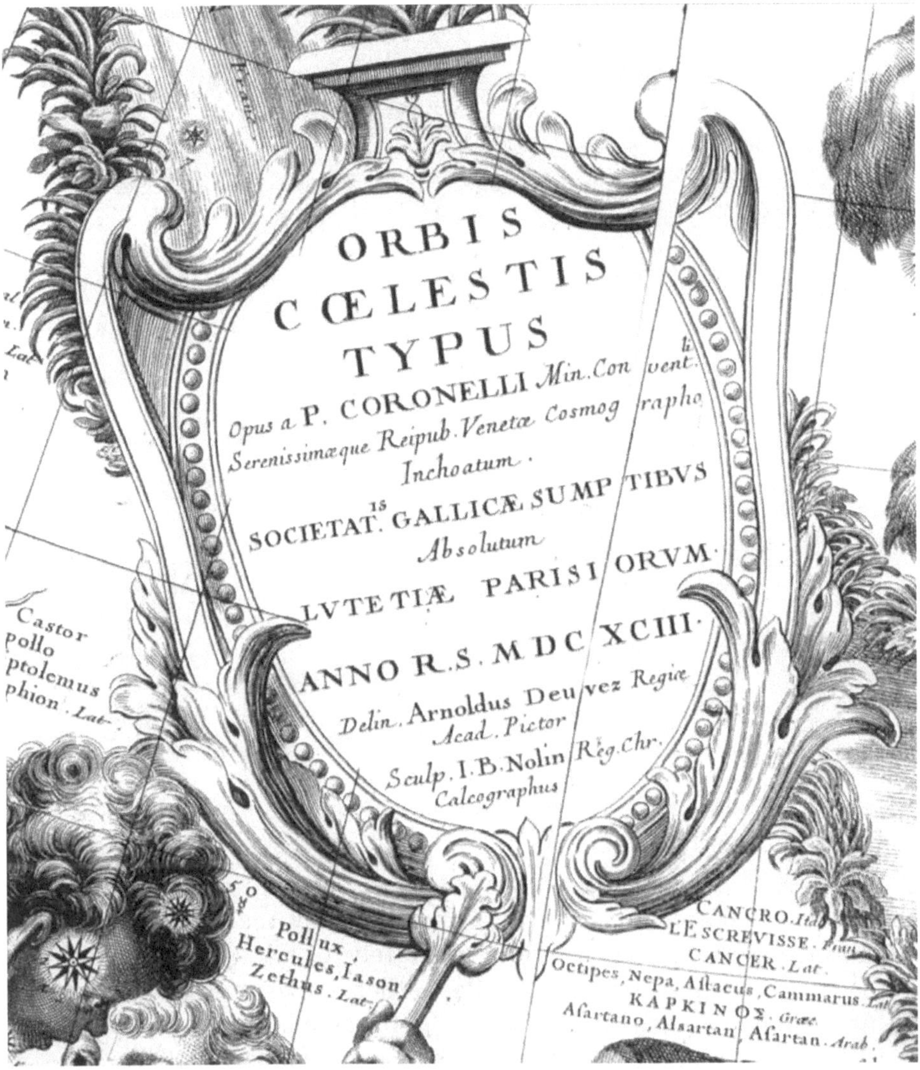

Figure 2. Cartouche (splitted over Northern Sectors 4 and 5) of the Coronelli 1693 Celestial Globe. Artists A. Deuvez and I.B. Nolin are associated to the work. (From the author's own prints obtained from the 1693 original plates; © Musée du Louvre)

Sciences. Stars and planets were positioned at their locations corresponding to the birth of Louis XIV, which induced the popular comment that the globes reproduced the King's horoscope.

Those globes are currently located at the National Library of France. If they were exceptional for their size (about four meters) and for the wealth of information they carried for the time, that very size made them cumbersome, difficult to appreciate and almost impossible to read. This explains

Figure 3. The same cartouche as the one depicted on Fig. 2, but barely decypherable on the Strasbourg Coronelli globe. (© A. Heck)

why they have been stored away in boxes for many, many years. The King had also to get the texts reproduced separately for the benefit of his guests.

Coronelli remained two years in Paris, then worked in various European places, including Paris again in August 1686 where he signed contracts with Jean-Baptiste Nolin (1657-1725) for etching (on copper) a 108cm celestial globe and other maps. In 1688, the globe saw a first release[3], with a second version in 1693. There were in total four editions for the terrestrial globe and six for the celestial one (two in Paris and four in Venice). They largely replaced in European libraries the smaller and older globes by Dutch Cartographer Willem Janszoon Blaeu (1571-1638).

The Strasbourg globe has been made with plates of the 1693 series as indicated by the cartouches displayed in Figs. 2 & 3. Coronelli's name and affiliation are followed by his title as *Cosmographer of the Republic of Venice* bestowed on 12 March 1685 by the Senate of the *Serenissima*. Etcher Nolin

[3]Together with a terrestrial globe etched in Venice. They were scaled-down versions of the globes for Louis XIV.

Figure 4. A large fragment is missing from the Strasbourg Coronelli globe: the fish representing the constellation *Doradus* in the lower part of Southern Sector 3, perhaps simply removed for a silly joke. (© A. Heck)

is also mentioned as well as Painter Arnould de Vuez (1644-1720), whose name was spelled there as Deuvez.

In 1684, Coronelli had founded in Venice what was perhaps the very first geographical society, the *Accademia Cosmographica degli Argonauti*. The purpose was before all to ensure a large distribution of his globes,

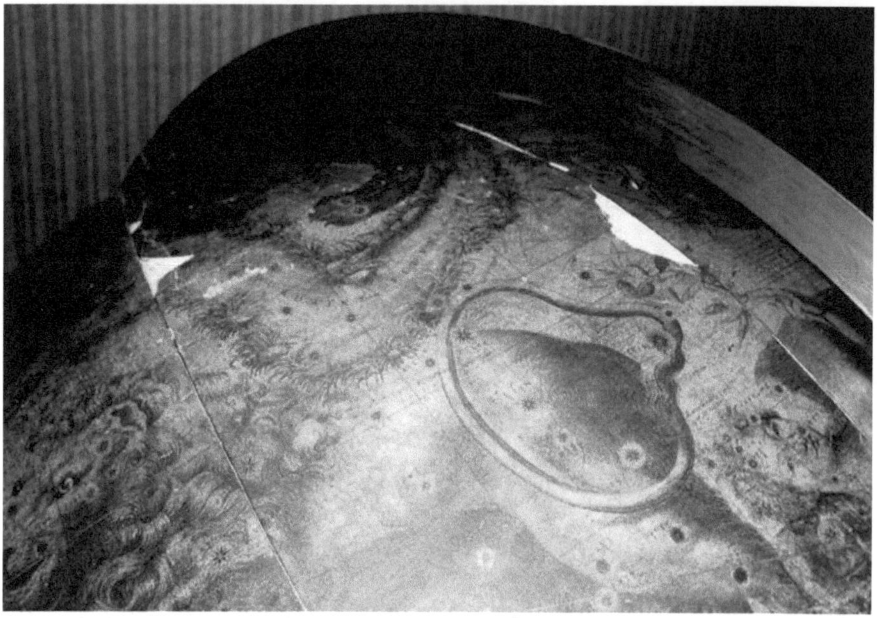

Figure 5. View of the Northern part of the Strasbourg Coronelli globe showing missing bits. (© A. Heck)

Figure 6. Vincenzo Coronelli (1650-1718).

maps and atlases. Pelletier & Roger (1993) list some of the eighty-four members of the society who were based in Paris (almost as many as in Venice), including Astronomer Jean-Dominique Cassini. The society was dissolved at the death of its founder.

Coronelli returned to Venice in 1705 where he stayed then permanently and died in 1718. Pelletier & Roger (1993) comment that excessiveness – be

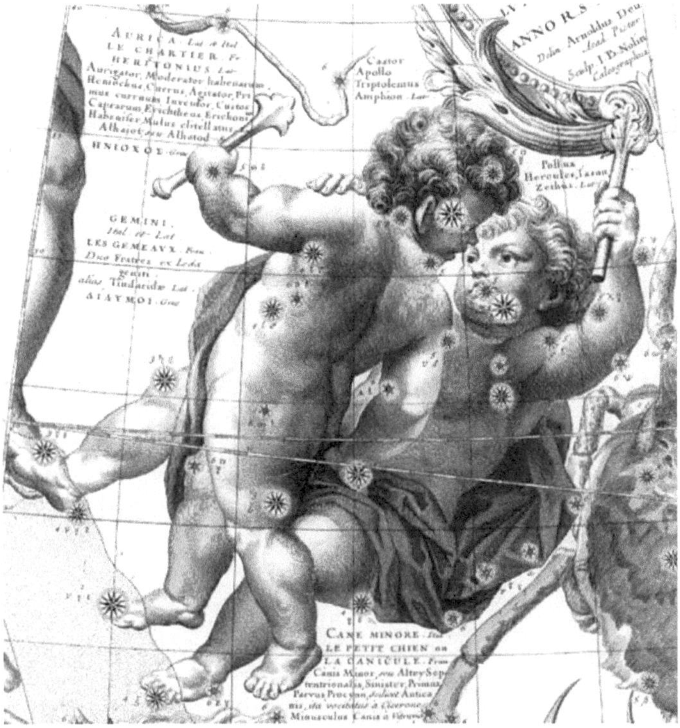

Figure 7. The Gemini Constellation on the Coronelli 1693 Celestial Globe, located near the cartouche described on Figs. 2 & 3. (From the author's prints by the 1693 original plates; © Musée du Louvre)

it for the size of globes[4] or the number of volumes published – was a constant feature of Coronelli's projects. This even led him to some difficulties with his religious order, especially about the induced expenses.

Coronelli's initiation to astronomy dated back to his first religious years. Colbert also requested him to test astronomical refractors.

3. More (but not everything yet) on the Strasbourg Globe

A question *not* to be asked is how many Coronelli globes have been produced since, just I did myself in the 1980s, prints from the original plates etched by Nolin can be ordered from the French Réunion des Musées Nationaux (RMN) [National Museums]. I have seen some of those prints, roughly colored, hanging in fast-food places.

A simple search through the web reveals a booming industry of Coronelli-labeled items, the authenticity of which is dubious (copies) and

[4]In 1708, he considered building globes four times bigger than those for Louis XIV.

Figure 8. Restored celestial (foreground) and terrestrial Coronelli globes from the Palais des Beaux-Arts in Lille. Note that here the axes are horizontal. (Picture A. Heck, by permission; © Palais des Beaux-Arts, Lille)

the exactitude of which has to be questioned. For unscrupulous merchants and unaware customers, Coronelli maps and globes, likewise Blaeu's, are nothing more than old-looking cartographic material. Away from all this, it is however interesting to attempt determining the trajectory of a specific genuine globe.

Figs. 2 & 7 have been reproduced from my own collection of prints obtained in the early 1980s from RMN's chalcographic workshop. The 26 original copper plates (12 Northern sectors, 12 Southern sectors and two polar caps) of the 1693 edition are stored there and were used for the printing. Fig. 7 in particular gives an idea of the fine etching work.

An idea of the magnificence of the original globes is for instance given by the two (celestial and terrestrial) exposed at the Palais des Beaux-Arts in Lille (Fig. 8), masterfully restored (Pelletier & Roger 1993). Note that here the axes are horizontal. The mountings as well as the stands were made along the wishes of the customers. As to the spheres, they were made of plaster covered by printed canvas.

Figure 9. The celestial globe from the *Société Industrielle de Mulhouse (SIM)* as displayed at the BUSIM library. (Picture A. Heck, by permission; © SIM, Mulhouse)

So far the history of the Strasbourg celestial globe or its trajectory through the centuries could not be traced. It is not mentioned in the (apparently exhaustive) *Inventar* of the first German period of the observatory (see Heck 2005). The fact that it appears first in a picture taken at the beginning of the 1920s could let one believe that it was brought by the French team just after WWI, but this is not certain at all. The globe might as well have been in the area beforehand. Some unsubstantiated rumors speak of a donation by a local scholar.

Another 1693 celestial globe (Fig. 9), with a mounting similar (although not fully identical) to the Strasbourg one, belongs to the Société Industrielle de Mulhouse (SIM) as a 1961 donation from a local wealthy family. It is in a quite good shape, with colors well visible. It underwent a restoration in 1982/83.

An investigation carried out in collaboration with SIM's David Carita revealed that the globe was acquired by the donator towards the end of 1928 from a gentleman called Frédéric Rossel who happened to advertize early that year (Rossel 1928) he had a Coronelli globe and several sets of prints that could be mounted in the same way as his own globe. Given the geographical proximity of Mulhouse and Strasbourg, was this pointing at a common origin for the two globes?

On the Strasbourg globe, the inclination of the polar axis can be adjusted at the local latitude by moving a supporting bronze circle on which the following inscription has been carved: *faict par Gatellier fabricateur des instrumens de mathematique 1697 Paris* [made by Gatellier manufacturer of instruments for mathematics 1697 Paris][5].

That mention indicates that the globe, in spite of being of the 1693 version, was probably not released before 1697-98, but it definitely dates back to the end of the 17th century. Duprat's (1970) list includes the globe, but without more details than its size and manufacturing date. No additional information could be found in the archives of the Département des Cartes et Plans at the Bibliothèque Nationale de France, nor from the local museums.

So the search for the successive owners/holders of the Strasbourg globe continues ...

Acknowledgments

Our warmest thanks go to Mmes & Messrs. M. Barnoud, B. Brejon de Lavergnée, D. Carita, N. Cramer, H. Richard, & Ph. Vonflie.

References

1. Duprat, G. 1970, Liste des Globes Terrestres et Célestes Anciens (antérieurs à 1850) Conservés dans les Collections Publiques de France, *Centre National de la Recherche Scientifique, Recherche Coopérative sur Programme* **172**.
2. Esclangon, E. 1926, La Nouvelle Organisation de l'Observatoire, *Ann. Obs. Strasbourg* **I**, 1-44.
3. Heck, A. 2005, Strasbourg Observatory and its Multinational History, this volume.
4. Pelletier, M. & Roger, A. 1993, La Renaissance des Globes de Coronelli (1650-1718) au Musée des Beaux-Arts de Lille, *Revue du Louvre* **4-1993**, 65-75.
5. Rossel, F. 1928, *Bull. Soc. Astron. France* **42**, 51.

[5]The French spelling would be today: *fait par Gatellier fabricant d'instruments pour les mathématiques.*

STRASBOURG GREEN RAYS

A. HECK
Observatoire Astronomique
11, rue de l'Université
F-67000 Strasbourg, France
heck@astro.u-strasbg.fr

Abstract. The literature dealing with the Cathedral of Strasbourg has known a recent boom thanks to a green-beam phenomenology taking place around the equinoxes within the building itself. After discussing this now fashionable popular attraction (without real significance), we detail Danjon & Rougier's first spectrographic observations of the solar green-ray phenomenon carried out in 1920 on the platform of the Cathedral. They confirmed the rôle of selective atmospheric absorption and explained the sometimes abnormal duration of the "flashes" as resulting from a combination between dispersion, turbulence and terrain irregularities at the horizon.

1. Strasbourg Cathedral Green Beam

Acknowledged as one of the most beautiful and most original achievements of gothic art, Strasbourg Cathedral is actually mixing several styles that have been masterfully integrated over the centuries. Initiated in 1015 along the roman style, its construction resumed in 1176 after a fire destroyed the first building.

The current facade displays triumphally the purest gothic of the 13th and 14th centuries, while the unique spire is from the 15th century (Fig. 1). Gigantic masterpiece of virtuosity, that building has been used diversely over time (as a church, a temple, a storage area, ...) and has been left almost untouched by the various conflicts in the region.

Since a few years, more and more tourists are attracted by a remarkable luminous phenomenon inside the Cathedral: at the equinoxes, around local Noon, a green beam scans the pulpit from West to East and passes over the

A. Heck (ed.), The Multinational History of Strasbourg Astronomical Observatory, 255–261.

head of Christ on the cross. Glossy volumes featuring the effect on colorful covers can be found in virtually all the city bookshops.

That exceptional green light is generated by sunrays passing through a piece of transparent green glass within one of the stained-glass windows. Intentional phenomenon ou pure coincidence? And what could be its interpretations?

It is easy to formulate these:
• an *astronomical* interpretation seeing in the green piece of glass and the vertical axis of the cross that it illuminates the elements of a equinoxial sundial signalling the season changes;
• an *artistic* interpretation since the pulpit, a jewel of flamboyant gothic, is singled out through that green light scan;
• a *religious* and *mystical* interpretation since the Spring equinox is close to Easter, Christ on the cross being then illuminated near his resurrection.

Is this then a demonstration of the genius of artists on stained glass, expressing their ability, their mystical and religious feelings while leaving a testimony to posterity? Such examples are available, generally centered on the solstices, as at the Chartres Cathedral. But nothing is less sure in Strasbourg. A detailed and well-documented sudy by Louis Tschaen (1986) tempers indeed the most imaginative enthusiasms.

An astronomical analysis of the phenomenon shows that the green beam passes over the head of Christ about one hour before true Noon (most elevated position locally of the Sun during the day) and this, one or two days after the Spring and the Fall equinoxes. Moreover, because of screening obstacles, the green light is visible from the pulpit axis only during approximately one month around the equinoxes. In terms of precision, much better results were already achievable at the time of the building construction.

If the stained-glass window from which the green beam originates dates back to 1875, the first mention of the phenomenon is much more recent, apparently from 1972 only. It is however another announcement in 1984 that had a real echo in the media and that caused the current touristic interest.

Other indications and detailed crosschecks by Tschaen (1986) converge towards a recent repair of the window (likely one of those carried out around 1950) that would have introduced a piece of glass of a different composition, allowing a well-contrasted beam to reach the pulpit.

Tschaen (1986) concludes to a high probability of an accidental nature of that equinoxial illumination, even if the various documents that would definitely prove it have not been found and might never be found.

Figure 1. Strasbourg Cathedral on an early 20th-century postcard. The platform level (66m elevation to ground) from which Danjon and Rougier carried out their observations in 1920 is clearly visible.

2. The Solar Green Rays

The solar green-light effects (rims, segments, rays or "flashes") are now widely known and extensive documentation can be found on them (see for instance Andrew T. Young's web site[1]). They can be observed at sunset (or just before sunrise) in favorable conditions of visibility: clear sky and far, unobstructed horizon.

[1] `http://mintaka.sdsu.edu/GF/`

Figure 2. Spectrograph installed in 1920 by Danjon and Rougier on the platform of Strasbourg Cathedral to study the green flash. The spire visible on the left of the picture is the *Temple Neuf* one while the faint structure perceptible to its right is *St Pierre le Jeune* spire. (© Obs. Astron. Strasbourg)

Physics easily explains the phenomenon through decomposition of sunlight by atmospheric refraction. Because of it, objects near the horizon appear higher in the sky than they actually are. Additionally, the refraction is more important for the short wavelengths (green, blue) than for the longer ones (yellow, orange, red). Thus near the horizon the green image (disk in the case of the Sun) is very slightly higher on the horizon than the red image (disk).

One must also take into account another phenomenon: the general absorption and dispersion of sunlight by the atmosphere, much more effective on the blue rays and *de facto* causing the blue background of the sky during the day. For the same reason, and because of the much thicker atmospheric layer crossed by the rays at sunset or sunrise (tangentially to Earth surface) than at Noon, the images of the Sun are then more reddish. Apart from local peculiarities, the sunset images are also generally redder than the sunrise ones because of an increased atmospheric absorption at the end of the day (water evaporation and increased aerosol turbulence during the day, etc.). All in all, this clarifies why the shortest solar wavelengths of interest here come from the green region of the spectrum.

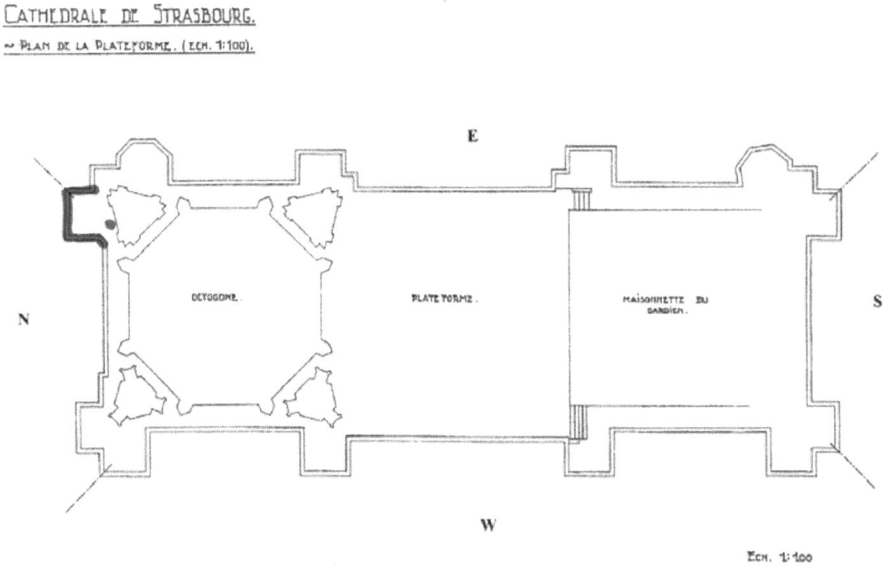

Figure 3. Sketch of Strasbourg Cathedral platform showing where Danjon & Rougier's spectrograph was located in 1920 to study the green flash: the Northeast corner (reinforced line) identified thanks to the spires discernible in the far on Fig. 2. (Courtesy and © Oeuvre Notre-Dame)

Thus, as the Sun gradually descends under the horizon, the slightly shifted disks corresponding to the red, orange, yellow, green parts of the spectrum disappear one after the other. Whether the witnesses will perceive the last disappearance as a *rim*, a *segment*, a *ray*, a *flash* or something else will depend, beyond unavoidable subjectivity, of the atmospheric conditions of the moment (layers, turbulence, ...) and of terrain irregularities at the horizon that will determine the actual shape of the green-light source (as seen from the observer) as well as the duration of the phenomenon. If the meteorological conditions and the othe relements are helping, a last green dot could appear as dissociated from the solar disk and bathe the observer in a fugitive emerald light – a magic and moving moment.

Extensive travelling has now allowed people not living in propicious areas to witness green-ray phenomena at sunsets here or there. Documentation is also easily available. This was far from being the case at the end of the 19th century or at the beginning of the 20th one. Then novels such as Jules Verne's *"Le Rayon Vert"* (1882) were about the only popularizing outlets, in the typical style of the time and only towards a relatively educated readership.

3. Danjon & Rougier's study of the 'Rayon Vert'

In line with the fashion of the time, André Danjon (1890-1967) and Gilbert Rougier (1886-1947) – who had arrived in 1919 at Strasbourg Observatory after the return of Alsace-Moselle to France (see *e.g.* Heck 2005) – got interested in the solar green-ray phenomenon. As sunsets and sunrises are not visible from Strasbourg Observatory itself, they installed in 1920 a spectrograph (Fig. 2) on the platform of Strasbourg Cathedral (Fig. 3) which gave them an elevated position (66m) over the surrounding ground and the plain of Alsace. The sunset takes place in Alsace behind the Vosges, but the observing period was chosen for having the Sun disappearing behind a rather low line. The observations were stopped when the Sun started setting more to the South, behind higher and higher summits.

The spectrograph was articulated around a heavy flint 60° prism, a Tessar-Krauss objective (f = 26cm) and a photographic chamber with a Zeiss objective (f = 52cm). The photographic plate holder was movable, allowing to record six spectra on 9cm×12cm panchromatic Wratten plates. The whole structure was mobile, allowing to track the Sun and to maintain the spectrograph slit horizontal and tangent to the solar edge. The exposures lasted six seconds for the center of the disk, one to two minutes for the fringes.

In total 19 spectrographic plates were obtained between 17 July and 3 August 1920, among which three spectra of the red fringe, four of the green light and ten of the general disk.

Danjon & Rougier (1920, but mainly 1926) made a clear distinction between the rim or fringe (*frange*) and the flash or ray (*rayon*), confirming the conclusions of earlier observers that the flash is the top of the fringe emerging alone during a short while. After correctly explaining the phenomenon through the combined effects of refraction and molecular diffusion (cf. Sect. 2), they entered a long discussion on the duration of the green ray, concluding that it should theoretically last about one second and a half but noting that much longer durations had been observed. Their second interrogation was why, while theoretically the rim should progressively change from red to green via orange and yellow, the visual observer was receiving without any transition a flash of pure green light.

Their spectroscopic observations allowed them to answer the last question by concluding that the solar spectrum should not be considered as a continuum, since most of the light was concentrated near the horizon in red and green Brewster bands making the sunsets (or sunrises) discontinuous phenomena.

As to the controversy on the duration of the green-light phenomenon, they comment that the *theoretical* duration varies rapidly with the elevation

Figure 4. Were those Alsacians (on an early 20th-century postcard) looking for the *green flash* from the Cathedral platform? Probably not since their shadows indicate they were turning their back to the Sun ...

of the horizon, the longest durations being observed from mountain sites since then the horizons are comparatively depressed. About actual durations of green phenomena reaching sometimes several seconds, they quite wisely conclude that nothing can be said without knowing the exact nature (slopes, obstacles, fragmentary horizon, etc.) of the terrain behind which the Sun is seen setting.

Acknowledgments

It is a real pleasure to acknowledge here the assistance of Ms. Sabine Bengel from *L'Oeuvre Notre-Dame*.

References

1. Danjon, A. & Rougier, G. 1920, Le spectre et la théorie du rayon vert, *C.R. Acad. Sciences Paris* **171**, 814-817.
2. Danjon, A. & Rougier, G. 1926, Le rayon vert – Étude spectroscopique et théorie, *Ann. Obs. Strasbourg* **I**, 105-115.
3. Heck, A. 2005, Strasbourg Astronomical Observatory and its Multinational History, this volume.
4. Tschaen, L. 1986, La lumière verte équinoxiale de la Cathédrale, *Bull. Cathédrale Strasbourg* **17**, 89-94.
5. Verne, J. 1882, Le rayon vert, Gauthier-Villars, Paris, 195 pp.

STRASBOURG OBSERVATORY COUNCIL MEMBERS

A. HECK & CH. BRUNEAU
Observatoire Astronomique
11, rue de l'Université
F-67000 Strasbourg, France
heck@astro.u-strasbg.fr
chantal@newb6.u-strasbg.fr

Abstract. The following note details, for the period 1969-2000, the elected members of the successive councils for Strasbourg Astronomical Observatory.

After the 1968 events that rocked European universities, the by-laws regulating French research institutions such as the observatories underwent several changes. Directors in particular were no more nominated in their position for the rest of their active life, but elected for limited terms. They became also answerable to a Council essentially made of elected members.

Table 1 gathers together, to the best of what could be extracted from the Observatory archives, the elected members of Strasbourg Observatory Council along the years (1969 to 2000).

The terms have been of three or four years, depending of the statutes ruling the place at the time. The 1982-1986 term has been extended pending the adoption of a decree organizing things anew. Some partial elections and replacements for various reasons[1] interfered with the general rythm.

For the sake of simplification, representatives of external organizations[2] (CNRS, INSU, Regional Council, etc.) as well as *de jure* members (such as the Directors when not elected members) have not been included.

The council under consideration is the official "Conseil d'U.E.R.[3]" (or equivalent) to which a fuzzy Observatory scientific council has sometimes

[1]Essentially resignation, change of electoral body, end of student status, ...
[2]Representatives from external scientific bodies would often change from one meeting to the next. Those from local political bodies rarely showed up more than once ...
[3]*U.E.R.* = *Unité d'Enseignement et de Recherche*, in other words, the lab.

A. Heck (ed.), The Multinational History of Strasbourg Astronomical Observatory, 263–269.

been amalgamated. The members of this latter council have however not been included either. Occasional invited attendees have been disregarded too.

Table 1 is followed by an alphabetical list.

TABLE 1. Strasbourg Observatory Council members.

1969-1971	
Bru, P.	Leher, B.
Florsch, A.	Schmitt, A.
Huss, J.	Traut, B.
Lacroute, P.	
1971-1973	
Fehrenbach, Ch.	Leher, B.
Florsch, A.	Traut, B.
Jung, J.	Valbousquet, A.
Lacroute, P.	
1973-1976	
Delhaye, J.	Jung, J.
Dubois, P.	Leher, B.
Fehrenbach, Ch.	Ochsenbein, Fr.
Florsch, A.	Traut, B.
1976-1979	
Delhaye, J.	Fresneau, A.
Egret, D.	Jaschek, C.
Fehrenbach, Ch.	Traut, B.
Florsch, A.	Wenger, M.
1979-1982	
Acker, A.	Florsch, A.
Bruneau, Ch.	Jaschek, C.
Delhaye, J.	Ochsenbein, Fr.
Fehrenbach, Ch.	Traut, B.
1982-1986/7	
Delhaye, J.	Halbwachs, J.L.
Dubois, P.	Jaschek, M.
Fehrenbach, Ch.	Marcout, J.
Florsch, A.	Wagner, M.J.

1987-1991	
Baglin, A.	Hamm, M./Marcout, J.
Didelon, P.	Hartley, J./Mischler, V./Becht, I.
Dubois, P.	Heck, A.
Egret, D./Jasniewicz, G.	Jaschek, M.
Fresneau, A./Florsch, A.	Kovalevsky, J.
Halbwachs, J.L.	Wenger, M.
1991-1995	
Acker, A.	Halbwachs, J.L.
Bonnarel, Fr.	Jasniewicz, G.
Bruneau, Ch.	Kovalevsky, J.
Colin-Souffrin, S.	Lery, Th./Lesteven, S./Lejeune, Th.
Dubois, P./Motch, Chr.	Marcout, J.
Egret, D.	Ochsenbein, Fr.
1995-1999	
Bartlett, J.	Jasniewicz, G.
Blanchard, A.	Lançon, A.
Bonnarel, Fr.	Lesteven, S.
Chereul, E./Picard, S.	Monier, R.
Hameury, J.M.	Puget, J.L.
Hangouët, J.Y.	Simon, G.
1999-2004	
Arnaud, M.	Lançon, A.
Bauer, P.	Motch, Chr.
Bulot, S./Michel, L.	Omont, A.
Egret, D.	Schnell, Fr.
Freire Ferrero, R.	Siebert, A.
Halbwachs, J.L.	Woelfel, F.

The following alphabetical list is mentioning the term(s) next to each name, plus possible specificities.

- Acker, A. (1979-1982)
(1991-1995)

- Arnaud, M. (1999-2004) (elected in Jan. 2000 as outsider)

- Baglin, A. (1987-1991) (elected in Jan. 1988 as outsider)

- Bartlett, J. (1995-1999)
- Bauer, P. (1999-2004)
- Becht, I. (1988-1991) (replacing Mischler from Sep. 1988 on)
- Blanchard, A. (1995-1999)
- Bonnarel, Fr. (1991-1995)
(1995-1999)
- Bru, P. (1969-1971)
- Bruneau, Ch. (1979-1982)
(1991-1995)
- Bulot, S. (1999-2000) (replaced in Dec. 2000 by Michel)
- Chereul, E. (1995-1997) (elected in Dec. 1995 as student)
- Collin-Souffrin, S. (1991-1995) (elected in Jan. 1992 as outsider)
- Delhaye, J. (1973-1976) (elected in Dec. 1973 as outsider)
(1976-1979) (elected in Dec. 1976 as outsider)
(1979-1982) (elected in Dec. 1979 as outsider)
(1982-1986/7) (elected in Jan. 1983 as outsider)
- Didelon, P. (1987-1991) (elected in Feb. 1988 as student)
- Dubois, P. (1973-1976)
(1982-1986/7)
(1987-1991)
(1991-1993) (replaced in Mar. 1993 by Motch)
- Egret, D. (1976-1979)
(1987-1991) (replaced in May 1988 by Jasniewicz)
(1991-1995)
(1999-2004)
- Fehrenbach, Ch. (1971-1973) (elected in Jan. 1971 as outsider)
(1973-1976) (elected in Dec. 1973 as outsider)
(1976-1979) (elected in Dec. 1976 as outsider)

(1979-1982) (elected in Dec. 1979 as outsider)
(1982-1986/7) (elected in Jan. 1983 as outsider)

- Florsch, A. (1969-1971)
  (1971-1973)
  (1973-1976)
  (1976-1979)
  (1979-1982)
  (1982-1987)
  (1988-1991) (replacing Fresneau from Nov. 1988 on)

- Freire Ferrero, R. (1999-2004)

- Fresneau, A. (1976-1979)
  (1987-1988) (replaced in Mar. 1988 by Florsch)

- Halbwachs, J.L. (1982-1986/7)
  (1987-1991)
  (1991-1995)
  (1999-2004)

- Hameury, J.M. (1995-1999)

- Hamm, M. (1987-1990) (replaced in Apr. 1990 by Marcout)

- Hangouët, J.Y. (1995-1999)

- Hartley, J. (1987-1988) (replaced in Mar. 1988 by Mischler)

- Heck, A. (1987-1991)

- Huss, J. (1969-1971)

- Jaschek, C. (1976-1979)
  (1979-1982)

- Jaschek, M. (1982-1986/7)
  (1987-1991)

- Jasniewicz, G. (1988-1991) (replacing Egret from May 1988 on)
  (1991-1995)
  (1995-1999)

- Jung, J. (1971-1973)
  (1973-1976)

- Kovalevsky, J. (1987-1991) (elected in Jan. 1988 as outsider)
  (1991-1995) (elected in Jan. 1992 as outsider)

- Lacroute, P. (1969-1971)
  (1971-1973)

- Lançon, A. (1995-1999)
  (1999-2004)

- Leher, B. (1969-1971)
  (1971-1973)
  (1973-1976)

- Lejeune, Th. (1993-1995) (elected in Dec. 1993 as student)

- Lery, Th. (1991-1992) (elected in Dec. 1991 as student)

- Lesteven, S. (1992-1993) (elected in Dec. 1992 as student)
  (1995-1999)

- Marcout, J. (1982-1986/7)
  (1990-1991) (replacing Hamm from May 1990 on)
  (1991-1995)

- Michel, L. (2000-2004) (replacing Bulot from Dec. 2000 on)

- Mischler, V. (1988) (replacing Hartley from Mar. 1988 on)
  (replaced in Jun. 1988 by Becht)

- Monier, R. (1995-1999)

- Motch, Chr. (1993-1995) (replacing Dubois from Mar. 1993 on)
  (1999-2004)

- Ochsenbein, Fr. (1973-1976)
  (1979-1982)
  (1991-1995)

- Omont, A. (1999-2004) (elected in Jan. 2000 as outsider)
- Piquard, S. (1997-1999) (elected in Dec. 1997 as student)
- Puget, J.L. (1995-1999) (elected in Jan. 1996 as outsider)
- Schmitt, A. (1969-1971)
- Schnell, Fr. (1999-2004)
- Siebert, A. (1999-2004) (elected in Dec. 1999 as student)
- Simon, G. (1995-1999) (elected in Jan. 1996 as outsider)
- Traut, B. (1969-1971)
(1971-1973)
(1973-1976)
(1976-1979)
(1979-1982)
- Valbousquet, A. (1971-1973)
- Wagner, M.J. (1982-1986/7)
- Wenger, M. (1976-1979)
(1987-1991)
- Woelfel, F. (1999-2004)

CDS COUNCIL MEMBERS

CH. BRUNEAU & A. HECK
Observatoire Astronomique
11, rue de l'Université
F-67000 Strasbourg, France
chantal@newb6.u-strasbg.fr
heck@astro.u-strasbg.fr

Abstract. This chapter details, for the period 1972-2000, the membership of the scientific councils for the *Stellar Data Center (CDS)*, later called *Strasbourg astronomical Data Center*, installed at Strasbourg Observatory.

The *Centre de Données Stellaires (CDS)*, subsequently called *Centre de Données astronomiques de Strasbourg*, has been installed in 1972 at Strasbourg Astronomical Observatory by the decision 1/72/D0 of the French *Institut National d'Astronomie et de Géophysique (INAG)*. See other chapters in this volume (Heck 2005a&b, Jung 2005) for more details.

From the beginning, a scientific council was part of its structures. The by-laws (Art. 4) detail its composition as being made of six French astronomers plus six foreign ones, serving terms in principle of three years. The first council has been set up by INAG Decision 2/72/D0 while the nomination of J. Jung for five years as the first CDS Director resulted from INAG Decision 3/72/D0 dated 31 May 1972. Table 1 gives the composition of the successive councils. It is followed by an alphabetical list.

Other members were subsequently accreted to the CDS councils, such as the CDS Directors or the Observatory Directors (when not designated members), or representatives from organizations such as INAG[1], the French *Centre National de la Recherche Scientifique (CNRS)* and *Centre National d'Études Spatiales (CNES)*, as well as the *Federation of Astronomical and Geophysical Services (FAGS)*. They are not listed here as those representatives often changed from one meeting to the other.

[1]Later called *Institut National des Sciences de l'Univers (INSU)*.

A. Heck (ed.), The Multinational History of Strasbourg Astronomical Observatory, 271–276.

TABLE 1. CDS Council members.

1972-1975	
Blaauw, A. (ESO)	Fricke, W. (ARI)
Bijaoui, A. (Nice Obs.)	Hauck, B. (Lausanne Univ.)
Boulon, J. (Paris Obs.)	Jung, J. (CDS)
Cayrel de Strobel, G. (Meudon Obs.)	Jaschek, C. (La Plata Obs.)
Delhaye, J. (Paris Obs.)	Larsson-Leander, G. (Lund Obs.)
Fehrenbach, Ch. (Haute Provence Obs.)	Murray, C.A. (Greenwich Obs.)
1976-1978	
Blaauw, A. (ESO)	Jung, J. (French Sc. Mission, Washington)
Delhaye, J. (Paris Obs.)	Lederle, T. (ARI)
Duncombe, R.L. (Texas Univ., Austin)	Meyer, Cl. (CERGA)
Foy, R. (Meudon Obs.)	Monnet, G. (Lyons Obs.)
Hack, M. (Trieste Obs.)	Spite, Fr. (Meudon Obs.)
Hauck, B. (Lausanne Univ.)	
1979-1981	
Bijaoui, A. (Nice Obs.)	Houziaux, L. (Mons Univ.)
Crézé, M. (Besançon Obs.)	Lederle, T. (ARI)
Dluzhnevskaya, O. (USSR Astron. Cl)	Lortet, M.C. (IAP)
Foy, R. (Meudon Obs.)	Macchetto, F.D. (STScI)
Hauck, B. (Lausanne Univ.)	Turon, C. (Meudon Obs.)
Heidmann, J. (Paris Obs.)	Westerhout, G. (US Naval Obs.)
1982-1984	
Bijaoui, A. (Nice Obs.)	Macchetto, F.D. (STScI)
Crézé, M. (Besançon Obs.)	Mermilliod, J.Cl. (Lausanne Univ.)
Delhaye, J. (Paris Obs.)	Monsonego, G. (CCSC)
Dluzhnevskaya, O. (USSR Astron. Cl)	Paturel, G. (Lyons Obs.)
Houziaux, L. (Mons Univ.)	Turon, C. (Meudon Obs.)
Lederle, T. (ARI)	Westerhout, G. (US Naval Obs.)
1985-1987	
Abalakin, V.K. (USSR Astron. Cl)	Jahreiss, H. (ARI)
Andersen, J. (Copenhagen Obs.)	Mathez, G. (Toulouse Obs.)
Andrillat, Y. (Haute Provence Obs.)	Mead, J. (NASA GSFC)
Benvenuti (ST-ECF)	Mermilliod, J.Cl. (Lausanne Univ.)
Delhaye, J. (Paris Obs.)	Ochsenbein, Fr. (CDS)
Gaillard, Cl. (CCSC)	Requième, Y. (Bordeaux Obs.)

1988-1990	
Alecian, G. (Meudon Obs.)	Kunth, D. (IAP)
Delhaye, J. (Paris Obs.)	Mathez, G. (Toulouse Obs.)
Gaillard, Cl. (CCSC)	Mead, J. (NASA GSFC)
Golay, M. (Geneva Obs.)	Perault, M. (ENS)
Grosbøl, P. (ESO)	Piskunov, A.E. (USSR Astron. Cl)
Jahreiss, H. (ARI)	Wesselius, P. (SRON)
1991-1993	
Albrecht, M. (ESO)	Perault, M. (ENS)
Davoust, E. (Toulouse Obs.)	Riegler, G. (NASA HQ)
Grenon, M. (Geneva Obs.)	Taylor, B. (ESTEC)
Guibert, J. (Paris Obs.)	Turon, C. (Paris Obs.)
Guilloteau, St. (IRAM)	Wesselius, P. (SRON)
Montmerle, Th. (CEA)	Wielen, R. (ARI)
1994-1996	
Albrecht, M. (ESO)	Pasian, F. (Trieste Obs.)
Davoust, E. (Toulouse Obs.)	Riegler, G. (NASA HQ)
Grenon, M. (Geneva Obs.)	Roques, J.P. (CESR)
Guibert, J. (Paris Obs.)	Taylor, B. (ESTEC)
Guilloteau, St. (IRAM)	Turon, C. (Paris Obs.)
Omont, A. (IAP)	Wielen, R. (ARI)
1997-1999	
Bastian, U. (ARI)	Lequeux, J. (Paris Obs.)
Bennett, K. (ESTEC)	Omont, A. (IAP)
Bredekamp, J. (NASA HQ)	Pasian, F. (Trieste Obs.)
Colin, J. (Bordeaux Obs.)	Quinn, P. (ESO)
Courvoisier, Th. (Geneva Obs.)	Roques, J.P. (CESR)
Feissel, M. (Paris Obs.)	Vigroux, L. (CEA)
2000-2003	
Atteia, J.L. (Midi-Pyrénées Obs.)	Feissel, M. (IGN)
Bennett, K. (ESTEC)	Grewing, M. (IRAM)
Bertout, C. (A&A)	Omont, A. (IAP)
Bredekamp, J. (NASA HQ)	Quinn, P. (ESO)
Castets, A. (Bordeaux Obs.)	Vigroux, L. (CEA)
Courvoisier, Th. (Geneva Obs.)	Wallace, P. (RAL)

- Abalakin, V.K. (1985-1987)
- Albrecht, M. (1991-1996)
- Alecian, G. (1988-1990)
- Andersen, J. (1985-1987)
- Andrillat, Y. (1985-1987)
- Atteia J.L. (2000-2003)
- Bastian, U. (1997-1999)
- Bennett, K. (1997-2003)
- Benvenuti, P. (1985-1987)
- Bertout, C. (2000-2003)
- Bijaoui, A. (1972-1974)
- " " (1979-1984)
- Blaauw, A. (1972-1978)
- Boulon, J. (1972-1974)
- Bredekamp, J. (1997-2004)
- Castets, A. (2000-2003)
- Cayrel de Strobel, G. (1972-74)
- Colin, J. (1997-1999)
- Courvoisier, Th. (1997-2003)
- Crézé, M. (1979-1984)
- Davoust, E. (1991-1996)
- Delhaye, J. (1972-1978
- " " (1982-1990)
- Dluzhnevskaya, O. (1979-1984)
- Duncombe, R.L. (1975-1978)
- Fehrenbach, Ch. (1972-1974)
- Feissel, M. (1997-2003)
- Foy, R. (1975-1981)
- Fricke, W. (1972-1974)
- Gaillard, Cl. (1985-1990)
- Golay, M. (1988-1990)
- Grenon, M. (1991-1996)
- Grewing, M. (2000-2003)
- Grosbøl, P. (1988-1990)
- Guibert, J. (1991-1996)
- Guilloteau, St. (1991-1996)
- Hack, M. (1975-1978)
- Hauck, B. (1972-1981)
- Heidmann, J. (1979-1981)
- Houziaux, L. (1979-1984)
- Jahreiss, H. (1985-1990)
- Jaschek, C. (1972-1974)
- Jung, J. (1972-1978)
- Kunth, D. (1988-1990)
- Larsson-Leander, G. (1972-74)
- Lederle, T. (1975-1984)
- Lequeux, J. (1997-1999)
- Lortet, M.C. (1979-1981)
- Macchetto, F.D. (1979-1984)
- Mathez, G. (1985-1990)
- Mead, J. (1985-1990)
- Mermilliod, J.Cl. (1982-1987)
- Meyer, Cl. (1975-1978)
- Monnet, G. (1975-1978)
- Monsonego, G. (1982-1984)
- Montmerle, Th. (1991-1993)
- Murray, C.A. (1972-1974)
- Ochsenbein, F. (1985-1987)
- Omont, A. (1994-2003)
- Pasian, F. (1994-1999)
- Paturel, G. (1982-1984)
- Perault, M. (1988-1993)
- Piskunov, A.E. (1988-1990)
- Quinn, P. (1997-2003)
- Requième, Y. (1985-1987)
- Riegler, G. (1991-1996)
- Roques, J.P. (1994-1999)
- Spite, F. (1975-1978)
- Taylor, B. (1991-1996)
- Turon, C. (1979-1984)
- " " (1991-1996)
- Vigroux, L. (1997-2003)
- Wallace, P. (2000-2003)
- Wesselius, P. (1988-1993)
- Westerhout, G. (1979-1984)
- Wielen, R. (1991-1996)

Figure 1. Attendees to one of the many CDS meetings (November 1984) gathering together members of the CDS Scientific Council and participants to the scientific session on the steps of the Big Dome of Strasbourg Observatory. From left to right, first row: D. Egret, M. Duflot, M. Mayor, J.Cl. Mermilliod, Y. Andrillat, Ch. Fehrenbach, L. Houziaux, A. Florsch; second row: W. Gliese, C. Turon, T. Lederle, C. Jaschek, J. Chauville, F. Ochsenbein, G. Westerhout; last rows: H. Jahreiss, A. Acker, A. Pédoussaut, P. Dubois, R. Burnage, O. Atanackovič, M. Holzer, J.L. Halbwachs, A. Heck. Three unidentified persons are mostly hidden. (© A. Heck)

References

1. Heck, A. 2005a, Strasbourg Astronomical Observatory and its Multinational History, this volume.
2. Heck, A. 2005b, Vistas into the CDS Genesis, this volume.
3. Jung, J. 2005, The Stellar Data Center: Origins and Early Beginnings (1972-1974), this volume.

STRASBOURG OBSERVATORY SCIENTIFIC PERSONNEL (FROM FOUNDATION TO ABOUT YEAR 2000)

B. TRAUT & A. HECK
Observatoire Astronomique
11, rue de l'Université
F-67000 Strasbourg, France
bernard.traut@evc.net
heck@astro.u-strasbg.fr

AND

H.W. DUERBECK
Vrije Universiteit Brussel
Pleinlaan 2
B-1050 Brussel, Belgium
hduerbec@vub.ac.be

Abstract. This chapter is providing an alphabetical list of Strasbourg scientific personnel from foundation up to Year 2000.

1. Intoduction

This chapter is providing an alphabetical list of Strasbourg *scientific* personnel for a period going from foundation up to Year 2000. It is provided *bona fide* and to the best of the information we could collect on scientists actually part of the staff (*i.e.* not those visiting nor temporarily associated). Living scientists have been contacted by e-mail in order to cross-check their data. Two persons, not quite part of the permanent staff (Couder, Rosenberg) have been included for exceptional profile and exceptional contributions to the Observatory.

The successive entries are listing, whenever available or known, the dates of birth (B) and decease (D), as well as the various positions held in relation with Strasbourg Observatory. Some positions are mentioned as linked to the "University". It was the Imperial University during the first German Period (up to 1918), then the Faculty of Sciences during the following French

A. Heck (ed.), The Multinational History of Strasbourg Astronomical Observatory, 277–292.

periods until 1970 when it became Strasbourg 1 University, equivalent to the *Université Louis Pasteur (ULP)*.

2. Personnel listing

- **ACKER, Agnès, née KELLER**
  - B: 28 Jan 1940, Thann (France)
  - 01 Oct 1964: joining as "Assistant Délégué, Strasbourg University
  - 01 Apr 1966: "Assistant Titulaire", Strasbourg University
  - 01 Oct 1975: "Maître-Assistant Titulaire", Strasbourg 1 University
  - 01 Apr 1978: "Maître-Assistant 1ère Classe", Strasbourg 1 University
  - 15 Jun 1981: Planetarium Director
  - 01 Feb 1983: "Professeur 2e Classe", Strasbourg 1 University
  - 01 Jan 1990: "Professeur 1ère Classe", Strasbourg 1 University

- **BACCHUS, Pierre**
  - B: 10 Jul 1923 (France)
  - 01 Oct 1946: joining as "Attaché de Recherche" CNRS
  - 01 Jan 1949: "Aide-Astronome Stagiaire"
  - 01 Jan 1951: "Aide-Astronome Titulaire"
  - 01 Apr 1956: "Astronome-Adjoint Faisant Fonction"
  - 01 Oct 1960: "Astronome-Adjoint Titulaire"
  - 01 Nov 1961: leaving for Lille University

- **BARLIER, François**
  - B: 02 Aug 1933 (France)
  - 01 Oct 1955: joining as "Chargé des Fonctions Assistant"
  - 01 Dec 1959: leaving for Paris Observatory

- **BARTLETT, James**
  - B: 25 Jun 1963, Mountain View (USA-CA)
  - 09 Feb 1994: joining as "Boursier Chateaubriand"
  - 09 Sep 1994: "Collaborateur Scientifique sur Contrat à Durée Déterminée"
  - 01 Sep 1995: "Astronome-Adjoint Stagiaire"
  - 01 Sep 1997: "Astronome-Adjoint Titulaire"
  - 01 Apr 1999: leaving for Midi-Pyrénées Observatory

- **BATY, Hubert**
  - B: 04 Sep 1963, Chauché (France)
  - 01 Oct 1992: joining as "Maître de Conférence", Strasbourg 1 University

- **BAUSCHINGER, Julius**
  - B: 28 Jan 1860, Fürth (Germany)

- D: 21 Jan 1934, Leipzig (Germany)
- 01 Apr 1909: joining as Professor at Strasbourg University, coming from the Astronomisches Rechen-Institut Berlin
- 1909 to 1919: Director
- 01 Aug 1919: expelled by French authorities

• **BECKER, Ernst [Emil Hugo]**
- B: 11 Aug 1843, Emmerich am Rhein (Germany)
- D: 06 Aug 1912, Freiburg (Germany)
- 01 Dec 1887: joining as Professor at Strasbourg University, coming from Gotha Observatory
- 1887 to 1909: Director and Professor at Strasbourg University
- 01 Apr 1909: retirement

• **BIENAYMÉ, Olivier**
- B: 09 Feb 1956, Versailles (France)
- 01 Oct 1994: joining as "Astronome-Adjoint 1ère Classe", coming from Besançon Observatory

• **BISCHOFF, Marguerite, née LUDWIG, div. WALTHER**
- B: 21 Jul 1922, Mulhouse (France)
- D: 2000, Strasbourg (France)
- 21 Sep 1959: joining as "Chargé des Fonctions Aide-Technique Principal"
- 23 Jun 1961: "Aide-Technique Principal Stagiaire"
- 12 Nov 1962: "Aide-Technique Titulaire"
- 04 Sep 1962: "Assistante d'Observatoire Stagiaire"
- 21 Sep 1964: "Assistante d'Observatoire Titulaire"
- 01 Oct 1981: "Aide-Astronome Titulaire"
- 26 Jul 1987: retirement

• **BISKE, Felix [Ladislaus]**
- B: 20 Nov 1874, Płonsk (Poland)
- D: [date and place unknown]
- 15 Feb 1905: joining as "2. Assistent"
- 31 Mar 1907: leaving

• **BLANCHARD, Alain**
- B: 18 Feb 1960, Dakar (Senegal)
- 01 Oct 1993: joining as "Professeur 2e Classe", Strasbourg 1 University, coming from Paris 7 University
- 30 Aug 2001: leaving for Toulouse 3 University

• **BOURION, Marthe Léonie**
– B: 13 Nov 1899, Laval sur Vologne (France)
– B: [date unknown], Paris (France)
– 25 Dec 1919: joining as "Calculatrice"
– 01 Jan 1924: "Assistante d'Observatoire"
– 01 Jan 1938: "Assistante 1ère Classe"
– 30 Sep 1965: retirement

• **BRU, Pierre [Louis]**
– B: 23 Feb 1912, Aurillac (France)
– D: 04 Sep 1987, Arpajon sur Cère (France)
– 01 Nov 1934: joining as "Assistant Auxiliaire"
– 16 Apr 1934: "Assistant Suppléant"
– 15 Oct 1938: "Assistant Titulaire"
– 15 Oct 1955: "Chargé de Fonction Aide-Astronome"
– 01 Oct 1960: "Aide-Astronome"
– 01 Mar 1977: retirement

• **CARNERA, Luigi**
– B: 14 Apr 1875, Triest (Austria) [now Trieste (Italy)]
– D: 30 Jul 1962, Firenze (Italy)
– 15 Oct 1902: joining as "Assistent",
coming from the Astrophysics Section of Heidelberg Observatory
– 30 Sep 1903: leaving for the International Latitude Station at Carloforte

• **COHN, Berthold**
– B: 19 Jan 1970, Rawitsch (Germany) [now Rawicz (Poland)]
– D: 19 May 1930, Strasbourg (France)
– 01 Oct 1903: joining as "Hilfsassistent"

• **COUDER, André [Joseph Alexandre]**
– B: 27 Nov 1897, Alençon (France)
– D: 16 Jan 1979, Bourg-la-Reine (France)
– 1925: "Stagiaire"
– 1926: leaving for Paris Observatory

• **CRÉZÉ, Michel**
– B: 21 Dec 1945, Angers (France)
– 01 Sep 1989: joining as "Astronome 2e Classe"
– 1990 to 1995: Director
– 1990 to 1995: CDS Director
– 01 Dec 1991: "Astronome 1ère Classe"

– 01 Sep 1995: leaving for Bretagne-Sud University (Vannes)

• **DANJON, André [Louis]**
– B: 06 Apr 1890, Caen (France)
– D: 21 Apr 1967, Suresnes (France)
– 01 Jan 1919: joining as "Aide-Astronome"
– 01 Jan 1929: "Astronome-Adjoint"
– 1929 to 1945: Director
– 15 May 1946: Honorary Professor, Strasbourg University (Clermont-Ferrand)
– 01 Apr 1945: leaving for Paris Observatory

• **DUBOIS, Pascal**
– B: 20 Mar 1942, Périgueux (France)
– 01 Nov 1966: joining as "Assistant Faisant Fonction"
– 05 Jul 1967: "Assistant Stagiaire"
– 05 Nov 1970: "Assistant Titulaire"
– 01 Jun 1980: "Aide-Astronome"
– 01 Jan 1989: "Astronome-Adjoint 1ère Classe"
– 01 Oct 1992: "Astronome 2e Classe"

• **EBELL, [Carl Wilhelm Ludwig] Martin**
– B: 29 Apr 1871, Neuruppin (Germany)
– D: 28 Nov 1944, Neuruppin (Germany)
– 01 Jan 1897: joining as "Assistent", coming from Berlin Observatory
– 31 Jan 1902: leaving for the Editorial Office of the *Astronomische Nachrichten* (Kiel)

• **EBERT, [Johannes Karl] Wilhelm**
– B: 12 Dec 1871, Leipzig (Germany)
– D: 20 Nov 1916 [World War I front]
– 01 Feb 1902: joining as "Assistent"
– 30 Sep 1902: leaving

• **EGRET, Daniel**
– B: 28 Aug 1951, Paris (France)
– 1973: "Assistant"
– 1982: "Aide-Astronome"
– 1986: "Astronome-Adjoint"
– 1987: "Administrateur Provisoire"
– 1988 to 1990: "Directeur-Adjoint"
– 1990: "Astronome 2e Classe"
– 1995 to 2000: Director

– 1998: "Astronome 1ère Classe"

• **ESCLANGON, Ernest Benjamin**
– B: 17 Mar 1876, Mison (France)
– D: 28 Jan 1954, Eyrenville (France)
– 01 Oct 1919: joining as Professor, Strasbourg University
– 1919 to 1929: Director
– 1929: leaving to Paris Observatory

• **FEHRENBACH, Charles [Max]**
– B: 29 Apr 1914, Straßburg (Germany) [now Strasbourg (France)]
– 01 Oct 1941: joining as "Chargé des Fonctions d'Aide-Astronome" (Observatory in Clermont-Ferrand)
– 1943: leaving for Haute-Provence Observatory (Deputy Director)

• **FLORSCH, Alphonse**
– B: 02 Dec 1929, Enchenberg (France)
– May 1956: joining as "Assistant"
– Jul 1962: "Aide-Astronome"
– 01 Oct 1973: "Astronome-Adjoint"
– 1976 to 1987: Director
– 01 Apr 1986: "Astronome 2e Classe"
– 01 Oct 1995: retirement

• **FREIRE FERRERO, Rubens**
– 01 Oct 1980: joining as "Maître-Assistant Stagiaire", Strasbourg 1 University
– 01 Oct 1983: "Maître-Assistant Titulaire", Strasbourg 1 University
– 01 Jan 1985: "Maitre de Conférence", Strasbourg 1 University

• **FRESNEAU, Alain**
– B: 05 Mar 1944, Strasbourg (France)
– 1963: joining as "Calculateur Temporaire"
– 01 Oct 1965: "Chargé des Fonctions d'Assistant"
– 01 May 1966: "Assistant d'Observatoire Stagiaire"
– 01 May 1967: "Assistant d'Observatoire Titulaire"
– 1967 to 1968: "Aide-Astronome" at Algiers Observatory (technical cooperation as military service)
– 01 Oct 1969: "Aide-Astronome"
– Oct 1970 to Dec 1971: "Wissenschaftlicher Angestellter", "Deutsches Geodätisches Forschungsinstitut", Munich
– 01 Feb 1980: "Astronome-Adjoint"
– Sept 1982 to Jun 1986: ESA Senior Astrometrist, Space Telescope Science Institute, Baltimore

– 01 Apr 1986: "Astronome 2e Classe"

• **GENOVA, Françoise**
– B: 22 May 1954, Bones (Algeria)
– 01 Oct 1993: joining as "Directeur de Recherche 2e Classe" CNRS, coming from CNES (Paris)
– 01 Oct 1995: CDS Director

• **GUERBER, Louis**
– B: 14 Jul 1920, Pionnat (France)
– 01 Jun 1945: joining as "Aide-Astronome"
– 01 Oct 1948: leaving for teaching at Collège de Barr

• **GUILLOUT, Patrick**
– B: 27 Feb 1965, St Germain des Fossés (France)
– 01 Sep 1992: joining as "Allocataire de Recherche", Strasbourg 1 University
– 01 Jan 1999: "Ingénieur Contractuel"
– 01 Sep 1999: "Astronome-Adjoint 2e Classe Stagiaire"
[– 01 Sep 2001: "Astronome-Adjoint 2e Classe Titulaire"]

• **HALBWACHS, Jean-Louis**
– B: 23 Dec 1955, Mulhouse (France)
– 01 Oct 1981: joining as "Assistant d'Observatoire Stagiaire"
– 01 Oct 1983: "Assistant d'Observatoire Titulaire"
– 01 Jul 1988: "Astronome-Adjoint 2e Classe Stagiaire"
– 01 Jan 1989: "Astronome-Adjoint 2e Classe Titulaire"
– 01 Oct 1990: "Astronome-Adjoint 1ère Classe"
– 01 Sep 1998: "Astronome 2e Classe"

• **HALM, Jacob [Karl Ernst]**
– B: 30 Nov 1866, Bingen (Germany)
– D: 17 Jul 1944, Stellenbosch (South Africa)
– 16 Apr 1889: joining as "Hilfsassistent"
– 01 Oct 1892: military service
– 01 Nov 1893: "Assistent"
– 31 Aug 1895: leaving for Edinburgh Observatory

• **HAMEURY, Jean-Marie**
– B: 21 Nov 1955, Paris (France)
– 01 Sep 1993: joining as "Directeur de Recherche 2e Classe" CNRS, coming from LAEC (Paris)
– 2000 to 2005: Director

• **HARTWIG, [Karl] Ernst [Albrecht]**
- B: 14 Jan 1851, Frankfurt am Main (Germany)
- D: 03 May 1923, Bamberg (Germany)
- 01 Jul 1874: joining as "Hilfsassistent"
- 01 Jul 1877: "Assistent"
- 31 Mar 1884: leaving for Dorpat Observatory

• **HASSENSTEIN, Walter [Georg]**
- B: 13 Sep 1883, Königsberg (Germany) [now Kaliningrad (Russia)]
- D: 06 Jan 1961, Potsdam (Germany)
- 01 Jan 1910: joining as "1. Assistent"
- 01 Sep 1916: called to army
- 1919: expelled by French authorities

• **HECK, André [Hubert Joseph]**
- B: 20 Sep 1946, Jalhay (Belgium)
- 01 Apr 1983: joining as "Astronome-Adjoint Associé", coming from ESA IUE Observatory at Vilspa (Madrid)
- 01 Oct 1986: "Astronome-Adjoint"
- 1988 to 1990: Director
- 01 Jan 1997: "Astronome 1ère Classe"

• **HELLERICH, Johannes**
- B: 11 Feb 1888, Hamburg (Germany)
- D: 30 May 1963, Münster (Germany)
- 01 Aug 1941: joining as Professor at Strasbourg University, coming from Hamburg Observatory
- 1941 to 1944: Director
- 1944: prisoner by French authorities in Saint-Sulpice-sur-Tarn (released in Feb 1946)

• **HEYVAERTS, Jean [François]**
- B: 17 Feb 1943, Paris (France)
- 01 Oct 1992: joining as "Professeur 1ère Classe", Strasbourg 1 University, coming from Paris 7 University
- 01 Oct 1995: "Professeur Classe Exceptionnelle", Strasbourg 1 University

• **HUSS, Joseph**
- B: 01 Feb 1906, Sulzbad (Germany) [now Soultz-les-Bains (France)]
- D: 26 Oct 1995, Strasbourg (France)
- 01 Oct 1929: joining as "Chargé des Fonctions d'Aide-Astronome", coming from Algiers Observatory
- 14 Oct 1930: "Assistant Titulaire"

- 01 Nov 1933: "Aide-Astronome"
- 30 Sep 1971: retirement

• **IBATA, Rodrigo**
- B: 05 Aug 1967, Woking (England)
- 01 Oct 2000: joining as "Chargé de Recherche 1e Classe" CNRS, coming from MPIA (Heidelberg)

• **JASCHEK, Carlos [Karl Otto Rüdiger]**
- B: 02 Feb 1926, Brieg (Germany) [now Brzeg (Poland)]
- D: 12 Apr 1999, Salamanca (Spain)
- 01 Oct 1974: joining as "Professeur Associé", Strasbourg 1 University, coming from Geneva University, on leave from La Plata Observatory
- 1975 to 1995: CDS Director
- 01 Oct 1978: "Professeur Titulaire", Strasbourg 1 University

• **JASCHEK, Mercedes, née CORVALÁN**
- B: 13 Apr 1926, Buenos Aires (Argentina)
- D: 21 Nov 1995, Salamanca (Spain)
- 1975: joining as "Astronome Adjoint Associé"
- 01 Oct 1983: "Astronome-Adjoint Stagiaire"
- 01 Oct 1985: "Astronome-Adjoint Titulaire"
- 30 Sep 1991: retirement

• **JASNIEWICZ, Gérard**
- B: 16 Apr 1952, Longwy (France)
- 01 Nov 1984: joining as "Vacataire", coming from Lausanne Institute of Astronomy
- 01 Jun 1987: "Boursier Société de Secours des Amis des Sciences"
- 01 Jul 1988: "Astronome-Adjoint Stagiaire"
- 01 Jul 1990: "Astronome-Adjoint Titulaire"
- 01 Sep 1995: "Astronome 2e Classe"
- 01 Sep 1996: leaving for GRAAL (Montpellier)

• **JOST, Ernst [Heinrich]**
- B: 18 Apr 1877, Hamburg (Germany)
- D: beg. 1945, Königsberg (Germany) [now Kaliningrad (Russia)]
- 01 Nov 1904: joining as "1. Assistent"
- 31 Dec 1909: leaving for Königsberg Observatory as "Observator"

• **JUNG, Jean**
- B: 15 Aug 1944, Beaune (France)
- 01 Oct 1971: joining "Astronome-Adjoint Stagiaire", coming from a CNRS position at Paris Observatory

– 19171 to 1975: CDS Director
– 01 Oct 1973: "Astronome-Adjoint Titulaire"
– 01 Feb 1975: leaving for IRIA (Paris)

• **KAUFMANN, August**
– B: [date and place unknown]
– D: [date and place unknown]
– 1884: joining as "Hilfsassistent"
– 1888: "2. Assistent"
– 15 Apr 1889: leaving for teaching in Altkirch

• **KOBOLD, Hermann [Albert]**
– B: 05 Aug 1858, Hannover (Germany)
– D: 11 Jun 1942, Kiel (Germany)
– 01 Apr 1886: joining as "Observator"
– 1886 to 1887: Director ad interim
– 1901: Professor, Strasbourg University
– 31 Mar 1902: leaving for Kiel Observatory

• **LACROUTE, Pierre [Alphonse Marie]**
– B: 12 Jan 1906, Dijon (France)
– D: 14 Jan 1993, La Verrière (France)
– 01 May 1946: joining as Professor, Strasbourg University, coming from Toulouse Observatory
– 1946 to 1976: Director
– 20 Jun 1976: Honorary Professor, Strasbourg 1 University
– 01 Oct 1976: retirement

• **LALLEMAND, André**
– B: 29 Sep 1904, Sirey (France)
– D: 24 Mar 1978, Paris (France)
– 01 Nov 1926: joining as "Assistant", coming from the Lycée of Haguenau
– 21 Jul 1928: "Aide Astronome"
– 01 Jan 1938: "Astronome-Adjoint"
– 01 Aug 1943: leaving for Paris Observatory

• **LANÇON, Ariane**
– B: 12 May 1968, Den Haag (Netherlands)
– 01 Sep 1995: joining as "Maître de Conférence Stagiaire", Strasbourg 1 University, coming from Mount Stromlo Observatory
– 01 Sep 1996: "Maître de Conférence Titulaire", Strasbourg 1 University

• **LAUCHER, Marguerite Marie, née GRASS**
– B: 28 Mar 1897, Schiltigheim (Germany) [now France]
– D: [place and date unknown]
– 01 Oct 1919: joining as "Calculatrice"
– 01 Jan 1924: "Assistante"
– 01 Oct 1949: retirement

• **LIEBMANN, Julius**
– B: 06 Jul 1880, Padang (Sumatra)
– D: 22 Aug 1914 [World War I front, Belgium]
– 01 Apr 1907: joining as "2. Assistent"
– 30 Sep 1913: leaving for Berlin Observatory

• **MARTIN, Paul [Emanuel Karl]**
– B: 01 Jan 1867, Gollub (Germany) [now Golub Dobrzyn (Poland)]
– D: [date and place unknown]
– 01 Oct 1892: joining as "2. Assistent", coming from Göttingen
– 01 Jun 1893: leaving for Göttingen Observatory

• **MATZDORFF, Martin**
– B: 08 Jul 1889, Thorn (Germany) [now Torún (Poland)]
– D: 02 Nov 1914, Ieper (Belgium) [World War I front]
– Early 1914: joining as "freiwilliger Hilfsarbeiter"

• **MEYER, Harry**
– B: 16 Dec 1877, Hannover (Germany)
– D: [place and date unknown]
– 01 Apr 1902: joining as "Assistent"
– 30 Sep 1902: leaving

• **MONIER, Richard**
– B: 17 Nov 1956, Annecy (France)
– 01 Oct 1994: joining as "Astronome-Adjoint 2e Classe Stagiaire", coming from ESA IUE Observatory at Vilspa (Madrid)
– 01 Oct 1996: "Astronome-Adjoint 2e Classe Titulaire"
– 01 Aug 1999: "Astronome-Adjoint 1ère Classe"

• **MOTCH, Christian**
– B: 18 Nov 1953, Les Pavillons sous Bois (France)
– 01 Aug 1992: joining as "Chargé de Recherche 2e Classe" CNRS, coming from MPE (Garching)
– 1996: "Directeur de Recherche 2e Classe" CNRS

- **MULLER, Paul**
- B: 17 Nov 1910, Lörchingen (Germany) [now Lorquin (France)]
- D: 09 Jul 2000, Nice (France)
- 01 Oct 1931: joining as "Assistant Faisant Fonction"
- 01 Oct 1934: "Assistant Titulaire"
- 01 Oct 1944: "Astronome-Adjoint"
- 1940 to 1945: prisoner of war
- 01 Apr 1956: leaving for Paris Observatory

- **NECKER, Karl [August Friedrich]**
- B: 26 Nov 1867, Berlin (Germany)
- D: 23 Dec 1897, Cairo (Egypt) [train accident]
- 01 Jun 1893: joining as "2. Assistent"
- 01 Sep 1895: "1. Assistent"
- 15 Mar 1897 – 01 Sep 1897: leave of absence (health)
- 01 Nov 1897: definitively leaving

- **OCHSENBEIN, François**
- B: 15 Dec 1947, Wolschwiller (France)
- 01 Oct 1971: joining as "Assistant"
- 01 Oct 1983: "Astronome-Adjoint"
- 01 Jan 1985 to 31 Dec 1990: European Southern Observatory, Garching (Germany)
- 01 Jan 1999: "Astronome 1ère Classe"

- **PAKULL, Manfred**
- B: 14 Feb 1947, Hamburg (Germany)
- 1990: joining as "Astronome 2e Classe Invité"
- 01 Dec 1992: "Directeur de Recherche 2e Classe" CNRS

- **PICHON, Christophe**
- B: 07 Mai 1968, Rennes (France)
- 01 Jan 1998: joining as "Chargé de Recherche Stagiaire 2e Classe" CNRS
- 01 Jul 1999: "Chargé de Recherche Titulaire 2e Classe" CNRS

- **PUMMERER, Paul**
- B: 14 Jun 1884, Regensburg (Germany)
- D: after 1930 [exact date and place unknown]
- Aug 1913: joining as "außeretatsmässiger Assistent"
- 01 Mar 1916: called to the army

- **REDLICH, Erich**
- B: 20 May 1883, Oberstreit (Germany) [now Poland]

– D: 11 Mar 1918, Blamont (France) [World War I front]
– 01 Oct 1903: joining as "Hilfsassistent"
– 01 Oct 1904: leaving for Kiel Observatory

- **ROSENBERG Hans Oswald**

– B: 18 May 1879, Berlin (Germany)
– D: 26 Jul 1940, Istanbul (Turkey)
– 1903: joining as "freiwilliger Hilfsarbeiter"
– 1907: leaving for Göttingen Observatory

- **ROUGIER, Gilbert Léon Laurent**

– B: 12 Sep 1886, La Mulatière (France)
– D: 10 Mar 1947, Floirac (France)
– 01 May 1920: joining as "Aide-Astronome"
– 01 Jan 1933: "Astronome-Adjoint"
– 01 Jan 1937: leaving for Bordeaux Observatory

- **SCHILLER, Karl**

– B: 27 Mar 1882, Köln (Germany)
– D: 26 Apr 1979, Frankfurt (Germany)
– 01 Oct 1903: joining as "Hilfsassistent"
– 01 Oct 1904: "2. Assistent"
– 01 Oct 1913: "Assistent"
– 1919: expelled by French authorities

- **SCHMITT, Alfred**

– B: 30 Nov 1907, Bust (Germany)
– D: 02 Apr 1975 (France)
– 01 Apr 1949: joining as "Aide-Astronome Titulaire", coming from Algiers Observatory Observatory
– 01 Oct 1955 to 09 Apr 1958: Director, Quito Observatory (Ecuador), on UNESCO assignment
– 01 Jul 1962: "Astronome-Adjoint"
– 01 Oct 1974: "Astronome-Adjoint Honoraire"
– 01 Oct 1973: retirement

- **SCHMITT, Odette, née BANCILHON**

– B: 22 Sep 1908 (France)
– 01 Jan 1950: joining as "Assistante"
– 02 Oct 1956 to 10 Apr 1958: "Aide-Astronome" at Quito Observatory (Ecuador)
– 01 Jul 1964: retirement

• **SCHUR, [Adolf Christian] Wilhelm**
– B: 15 Apr 1846, Altona (Denmark) [now Germany]
– D: 01 Jul 1901, Göttingen (Germany)
– 01 Jul 1873: joining as "Assistent"
– 01 Jul 1877: "Observator"
– 1882 to 1886: Director ad interim
– 1886: leaving for Göttingen Observatory

• **SOBCZAC, Cesette Jeanne, née RIS**
– B: 03 May 1905
– D: [date unkown], Strasbourg (France)
– 01 Oct 1925: joining as "Calculatrice"
– 01 Jan 1968: "Assistante d'Observatoire"
– 31 Mar 1968: retirement

• **STECHERT, Carl [Friedrich Gottlieb Peter Heinrich]**
– B: 21 Aug 1856, Wismar (Germany)
– D: 06 Jun 1925, Hamburg (Germany)
– 15 Feb 1885: joining as "Assistent"
– 15 Nov 1885: leaving for the Astronomisches Recheninstitut Berlin

• **TETENS, Otto [Peter Harens]**
– B: 26 Sep 1865, Rendsburg (Germany)
– D: 15 Feb 1945, Teplitz-Schönau (Germany)
– 01 Feb 1898: joining as "Assistent"
– 15 Jan 1902: leaving for mounting a geophysical station in Samoa

• **VALBOUSQUET, Armand**
– B: 05 Aug 1942, Périgueux (France)
– 01 Aug 1966: joining as "Chargé des Fonctions d'Assistant"
– 06 Jan 1968: "Assistant Stagiaire"
– 06 Jan 1970: "Assistant Titulaire"
– 01 Oct 1973: "Aide-Astronome Titulaire"
– 01 Apr 1986: "Astronome-Adjoint 1ère Classe"
[– 2002: retirement]

• **VALLS-GABAUD, David**
– B: 08 Feb 1963, Madrid (Spain)
– 01 Oct 1994: joining as "Chargé de Recherches Stagiaire de 1ère Classe" CNRS
– 01 Jun 1996: "Chargé de Recherche Titulaire de 1ère Classe" CNRS

– 01 Sep 1998: leaving for Midi-Pyrénées Observatory (Toulouse)

- **VÉRONNET, Alexandre**

– B: 11 May 1876, Chagny (France)
– D: 21 Aug 1951 Chassey-le-Camp (France)
– 01 Oct 1919: joining as "Astronome-Adjoint", coming from Paris Observatory
– 30 Sep 1941: retirement

- **WALCH, Jean-Jacques**

– B: 02 Mar 1939 (France)
– 01 Dec 1966: joining as "Assistant Stagiaire"
– 05 Jul 1969: "Assistant Titulaire"
– 01 Oct 1969: "Aide-Astronome"
– 01 Apr 1975: leaving for CERGA (Grasse)

- **WANACH, [Carl] Bernhard**

– B: 11 Jun 1867, Mesothen (Germany) (now Mezotne (Estonia)]
– D: 02 Apr 1928, Potsdam (Germany)
– 01 Jul 1892: joining as "Assistent", coming from Pulkovo
– 31 Dec 1896: leaving for the International Geodetic Bureau (Potsdam)

- **WINNECKE, August [Friedrich Theodor]**

– B: 05 Feb 1835, Hildesheim (Germany)
– D: 02 Dec 1897, Bonn (Germany)
– 1872 to 1886: (first) Director and Professor of Astronomy at Strasbourg University
– Feb 1882: illness prevents him from continuing his work
– 31 May 1886: Professor Emeritus, Strasbourg University [retirement]

- **WIRTZ, Carl [Wilhelm]**

– B: 24 Aug 1874, Krefeld (Germany)
– D: 19 Feb 1939, Hamburg (Germany)
– 01 Apr 1902: joining as "Observator"
– 1909: "Professor", Strasbourg University
– 01 Sep 1916: called to army
– 06 Jan 1919: expelled by French authorities

- **WISLICENUS, Walter [Friedrich]**

– B: 05 Nov 1859, Haberstadt (Germany)
– D: 03 Oct 1905, Straßburg (Germany) [now Strasbourg (France)]
– Apr 1883: joining as "Assistent"
– 01 Oct 1889: "Privatdocent", Strasbourg University

- 01 Oct 1894: "Professor", Strasbourg University

• **ZWINK, Max [Arthur]**
- B: 21 Sep 1858, Waldenburg (Germany)
- D: after 1930 [exact date and place unknown]
- 01 Nov 1890: joining as "1. Assistent", coming from Berlin Observatory
- 01 Jul 1892: leaving for the Imperial Patent Office (Berlin)

Acknowledgements

We gratefully acknowledge the assistance of all the persons who helped us compiling this list and, in particular, those listed above or their relatives who answered promptly to our direct queries. We thank particularly Ms. R. Lukas-Rogalka (Staatsbibliothek Berlin) for supplying information from old PhD theses, as well as Mrs. Chr. Douyère-Demeulenaere (French National Archives) and the archivists of the French Academy of Sciences under the leadership of Mrs. Fl. Greffe.

STRASBOURG OBSERVATORY INSTITUTIONAL PUBLICATIONS

PH. VONFLIE & A. HECK
Observatoire Astronomique
11 rue de l'Université
F-67000 Strasbourg, France
philippe.vonflie@newb6.u-strasbg.fr
heck@astro.u-strasbg.fr

Abstract. This chapter details the publications produced by Strasbourg Astronomical Observatory as an institution, during both German and French times, up to about Year 2000.

1. Introduction

The institutional publications of Strasbourg Astronomical Observatory are detailed below for both the German and French periods[1]. Those publications have appeared under different series as explained hereafter.

Except for a few cases pointed out below, the papers published by individual scientists from the Observatory in the professional journals have not been included in this review. They should be looked for in the different volumes of the *Astronomischer Jashresbericht* [2] and of its continuation as *Astronomy and Astrophysics Abstracts* for instance, or in a specific database such as the one maintained by the *Astrophysics Data System (ADS)* [3].

Unless otherwise stated, all publications are under the current A4 format or a similar one (such as the quarto format for the earlier publications).

2. Annals of Strasbourg Observatory

This section details the first institutional publications of the Observatory published under the titles *Annalen der Kaiserlichen Universitäts-*

[1]See *e.g.* the introductory chapter by Heck 2005a in this volume.
[2]Actually conceived in Strasbourg – see Duerbeck 2005b in this volume.
[3]http://adsabs.harvard.edu/

A. Heck (ed.), The Multinational History of Strasbourg Astronomical Observatory, 293–310.

Sternwarte in Strassburg for the first German period and *Annales de l'Observatoire de Strasbourg* for the subsequent French periods.

2.1. ANNALEN DER KAISERLICHEN UNIVERSITÄTS-STERNWARTE IN STRASSBURG

The successive volumes are detailed hereafter.

2.1.1. *Erster Band*

Here are the contents of the first volume, published in 1896 by the Director E. Becker and printed in Karlsruhe by the G. Braun'sche Hofbuchhandlung:

- Der Bau und die Einrichtung der Sternwarte (pp. III-X)
- Die Lage der neuen Sternwarte (pp. X-XI)
- Der Repsold'sche Meridiankreis (pp. XII-XVII)
- Beobachtungen am Repsold'schen Meridiankreis 1882 März 15 – 1886 Sept. 9. Untersuchung des Instruments und seiner Aufstellung (pp. XXI-XCVI)
- Erläuterungen zu den Beobachtungen (pp. XCVII-XCVIII)
- Beobachtungen (pp. 1-319)
- Verbesserte Ablesungen der meteorologischen Instrumente (pp. 321-336)
- Temperatur in dem Fundamentenraum des Meridianintruments (pp. 337-338)
- Beiträge zu dem Klima von Strassburg nach den meteorologischen Beobachtungen von 1873-1879 (pp. 338-340)

2.1.2. *Zweiter Band*

Here are the contents of the second volume, published in 1899 by the Director E. Becker and printed in Karlsruhe by the G. Braun'sche Hofbuchhandlung:

- Beobachtungen am Repsold'schen Meridiankreis 1886 Sept. – 1888 April (pp. I-XX)
- Beobachtungen (pp. 1-154)
- Ableitung der definitiven Oerter der in den Jahren 1882-1888 beobachteten Fixsterne (pp. (I)-(XXV))
- Einzel-Resultate der Sternbeobachtungen 1882-1883 (pp. (1)-(6))
- Einzel-Resultate der Sternbeobachtungen 1884-1888 Abth. A (pp. (7)-(113))
- Einzel-Resultate der Sternbeobachtungen 1884-1888 Abth. B (pp. (115)-(145))
- I. Katalog von 254 Sternen für das Aequinoctium 1885 nach Beobachtungen in den Jahren 1882-1883 (pp. (147)-(153))

- II. Katalog von 858 Sternen für das Aequinoctium 1885 nach Beobachtungen in den Jahren 1884-1888 (pp. (155)-(174))
- III. Katalog von Correctionen von 368 Fundamentalsternen nach Beobachtungen in den Jahren 1884-1888 (pp. (175)-(184))
- Annex A:
  Resultate aus den an der Kais. Univ.-Sternwarte zu Strassburg angestellten Heliometer- und Refractormessungen der partiellen Sonnenfinsternisse 1890 Juni 16-17, 1891 Juni 6 und 1893 April 16 und aus allen bekannt gewordenen Contactbeobachtungen, abgeleitet von Dr. H. Kobold, Observator der Sternwarte (pp. A1-A39)
- Annex B:
  Ueber die Bestimmung der Form der Zapfen eines Durchgangsinstruments mittelst eines Axencollimators mit Anwendung auf den Repsold'schen Meridiankreis der Kais. Univ.-Sternwarte in Strassburg von Bernhard Wanach, Assistent der Sternwarte (pp. B1-B16)
- Annex C: Tafeln zur Berechnung der Praecession (p. I)
  Einleitung (p. III)
  Tafel I zur Berechnung der jährlichen Praecession (pp. C1-C73)
  Tafel II zur Berechnung der Variatio saecularis (pp. C75-C87)
  Tafel III zur Berechnung des dritten Gliedes der Praecession (pp. C89-C91)

2.1.3. *Dritter Band*

Here are the contents of the third volume, published in 1909 by the Director E. Becker and printed in Karlsruhe by the G. Braun'sche Hofbuchdruckerei:

- Beobachtungen von Nebelflecken, ausgeführt in den Jahren 1875-1880 am 6-zölligen Refraktor der provisorischen Sternwarte von A. Winnecke, bearbeitet von E. Becker (pp. VII-XIX, 1-209)
- Beobachtungen von Nebelflecken, ausgeführt in den Jahren 1880-1902 am 18-zölligen Refraktor der neuen Sternwarte von H. Kobold, A. Winnecke und W. Schur, bearbeitet von C.W. Wirtz (pp. (1)-(145))
  I. Teil 1880 bis 1896 Oktober im wesentlichen beobachtet von H. Kobold (pp. (1)-(72))
  II. Teil 1896 November bis 1902 März beobachtet von H. Kobold (pp. (75)-(145))
- Trabantenbeobachtungen in den Jahren 1881-1896, im wesentlichen ausgeführt von H. Kobold, bearbeitet von C.W. Wirtz (pp. [1]-[8])
- Über die Schraubenfehler der beiden Mikrometer des großen Refraktors (pp. [9]-[11])
- Annex A:
  Resultate der Beobachtung der Sonnenfinsternis 1900 Mai 28 nebst

einer Bestimmung der Reduktionselemente für das Heliometer von H. Kobold (pp. A1-A36)
- Annex B:
  Resultate der Beobachtungen der Mondfinsternisse 1888 Januar 28 und 1892 Mai 11 von H. Kobold (pp. B1-B11)
- Annex C:
  Resultate aus Beobachtungen während der Mondfinsternis 1903 April 11 von C.W. Wirtz (pp. C1-C18)

2.1.4. *Vierter Band – Erster Teil*
Here are the contents of the first part of the fourth volume, published in 1911 by the Director J. Bauschinger and printed in Karlsruhe by the G. Braun'sche Hofbuchdruckerei:

- Beobachtungen von Nebelflecken am 49cm Refraktor III. Teil 1902 April bis 1910 März von C. Wirtz (pp. 1-78)
- Generalkatalog der am großen Refraktor in Straßburg beobachteten Nebelflecke 1881-1910 (pp. 79-112)
- Vergleichung des Generalkataloges mit anderen Nebelverzeichnissen von C. Wirtz (pp. 113-176)

2.1.5. *Vierter Band – Zweiter Teil*
Here are the contents of the second part of the fourth volume, published in 1912 by the Director J. Bauschinger and printed in Karlsruhe by the G. Braun'sche Hofbuchdruckerei:

- Beobachtungen von Doppel- und Nachbarsternen am 49cm-Refraktor I. Teil 1902-1910 von Carl Wirtz (pp. 177-242)
- Beobachtungen der großen Planeten von Carl Wirtz (pp. 243-286)
- Beobachtungen von Satelliten von Carl Wirtz (pp. 287-303)
- Zusätze zu den Arbeiten über Nebelflecke im ersten Teil nebst einem "Versuch zur kosmischen Stellung der Nebel" von Carl Wirtz (pp. 304-316)
- Vergleichung des AG Kataloges Straßburg mit anderen Sternkatalogen von Berthold Cohn (pp. 317-431)

2.1.6. *Fünfter Band*
Here are the contents of the fifth volume, published in 1923 by J. Bauschinger and printed in Karlsruhe by the G. Braun'sche Hofbuchdruckerei:

- Beobachtungen am Repsold'schen Meridiankreis 1888 Mai 23 – 1893 Februar 8

Ableitung der definitiven Örter der in den Jahren 1888-1893 am Repsold'schen Meridiankreis beobachteten Fixsterne (pp. A1-A40)
Katalog der Korrectionen von 364 Fundamentalsternen (pp. A23-A28)
Katalog von 760 Sternen für das Äquinoktium 1890.0 (pp. A29-A40)

- Beobachtungen von Doppelsternen am 49-cm-Refraktor. II. Teil 1910-1914 (pp. B1-B79)

The following complement to the fifth volume has been published in 1926 by J. Bauschinger "in the name of the Strasbourg scientific society in Heidelberg". The printer was G. Braun GmbH vorm. G. Braun'sche Hofbuchdruckerei.

- Beobachtungen der Sonne, des Mondes und der großen Planeten am Repsold'schen Meridiankreis der Straßburger Sternwarte in den Jahren 1882-1893 (pp. C1-C52)

It should be noticed that these issues were published well after World War I (WWI), but are dealing with observations carried out before the conflict. Bauschinger revised them before final publication. Note also that the first volume of the French *Annales* described herafter contains a contribution from a German employee of Polish origin (Cohn[4]) who also authored a section in the fourth volume of the German *Annalen.*

2.2. ANNALES DE L'OBSERVATOIRE DE STRASBOURG

The following six volumes constitute what is called the 2nd Series of Strasbourg Observatory Annals. Thus a volume of Annals in French was published in Paris the same year (1926) that another one in German was printed in Heildelberg (see above).

2.2.1. *Tome I*

The first volume was published in 1926 by the then Director E. Esclangon. It were printed in Paris by Gauthier-Villars, printers-publishers for the Bureau des Longitudes. Note that authors were not always stated.

- Première partie:
  La nouvelle organisation de l'observatoire (pp. 1-58)
  Modifications apportées à l'outillage scientifique (pp. 59-100)
- Deuxième partie: Mémoires
  Le rayon vert – Étude spectroscopique et théorie, par A. Danjon et G. Rougier (pp. 105-118)
  La dispersion atmosphérique et la scintillation, par A. Danjon (pp.

[4]See Duerbeck (2005a) for more on Berthold (Baruch) Cohn, one of the longest employees at Strasbourg Observatory (1900-1930). According to the peace treaty of 1918, since he came from (then) Poland, he was not expulsed (Esclangon 1926).

119-128)
Les étoiles géantes. Constitution physique et évolution, par A. Véronnet (pp. 129-174)
Étude de l'objectif du grand réfracteur, par A. Danjon (pp. 175-188)
Les mouvements propres des étoiles du catalogue AG de Strasbourg, par M. Cohn[5] (pp. 189-291)

- Troisième partie: Observations
  Indications générales (p. 297)
  Tour de vis des micromètres des équatoriaux (p. 298)
  Mesures d'étoiles doubles, par G. Rougier (pp. 299-304)
  Observations de comètes
  Diamètre apparent de Vénus (pp. 307-308)
  Observations des anneaux de Saturne en 1921 (pp. 309-312)
  Occultations d'étoiles par la Lune (p. 313)
  Occultation d'étoile par Jupiter (p. 314)
  Observations des satellites de Jupiter (p. 315)
  Observations d'éclipses du Soleil (pp. 316-317)
  Observations photométriques de Nova Cygni 1920, par A. Danjon (pp. 318-320)
  Mesures micrométriques et photométriques de Cygni 1920, par G. Rougier (pp. 321-326)
  Détermination de l'intensité de la pesanteur à l'Observatoire de Strasbourg, par E. Esclangon (pp. 327-372)
  Sur la précision des observations méridiennes et des déterminations de longitude, par E. Esclangon (pp. 372-404)

2.2.2. *Tome II*

The second volume gathers together three sections published respectively in 1928, 1932 and 1933. E. Esclangon was still Director when the first section appeared, but A. Danjon had succeeded him for the two subsequent ones. Therefore the second volume has to be considered under their joint editorship. The printer was Henri Tessier in Orléans.

- Fascicule 1 (1928):
  Recherches de photométrie astronomique, par A. Danjon
  Principes et instruments (pp. 1-89)
  Observations et discussion des résultats (pp. 93-185)
- Fascicule 2 (1932):
  Observations d'Eros (1930-1931), par MM. Danjon, Rougier, Lallemand, Tremblot, Huss, Grouitch (pp. 187-200)

[5]Cohn also contributed to the fourth volume of the German *Annalen*. See Footnote 4.

- Fascicule 3 (1933):
  Photométrie photoélectrique globale de la Lune, par G. Rougier (pp. 205-339)

2.2.3. *Tome III*

The third volume gathers together five sections published over the years 1933 to 1937 with A. Danjon as Director of the Observatory. Only the first section is explicitely dated though. The printer was again Henri Tessier in Orléans.

- Fascicule 1 (1933):
  Réduction et discussion des occultations d'étoiles par la Lune, par V. Grouitch (pp. 1-111)
- Fascicule 2 (1936):
  L'objectif du grand réfracteur de Strasbourg, par A. Couder (pp. 1-134)
  Note sur le réglage du compensateur de trempe de M. A. Couder, par A. Danjon (pp. 135-138)
- Fascicule 3 (1936):
  Nouvelles recherches sur la photométrie de la lumière cendrée et l'albédo de la terre, par A. Danjon (pp. 139-180)
- Fascicule 4 (1936):
  Description et théorie d'un micromètre interférentiel à demi-onde, par A. Danjon (pp. 181-255)
- Fascicule 5 (1937):
  Comparaison photométrique de la Lune et du Soleil.
  Albédo photoélectrique de la Lune, par G. Rougier (pp. 257-282)

2.2.4. *Tome IV*

The fourth volume gathers together three sections published respectively in 1950, 1956 and 1959 under the editorship of P. Lacroute who had become the first post-WWII Director of the Observatory. The printer was J. & R. Sennac in Paris.

- Fascicule 1 (1950):
  Catalogue de 2251 étoiles faibles réduites à 1930.0 (pp. 5-62)
- Fascicule 2 (1956):
  Étude de quelques propriétés du cercle méridien, par P. Muller, P. Lacroute, A. Schmitt, J. Huss, P. Bru (pp. 5-15)
  Nouveau procédé de lecture de l'appoint en déclinaison dans les observations méridiennes, par P. Bacchus (pp. 17-23)
  Observation du Soleil en ascension droite, par P. Bacchus (pp. 24-27)

- Fascicule 3 (1959):
  Méthode photoélectrique d'observation des étoiles doubles, par P. Bacchus (pp. 5-67)

That fourth volume has been published after the fifth one hereafter, but it has been given that rank in order to respect the chronology of the observations carried out. As reminded by Lacroute in the first section, some data go back to 1928 and mainly to the range 1931-1934. They could not be published during WWII. As it happened earlier with Bauschinger and pre-WWI observations, the publication took place much after the conflict, after corrections and new calculations.

Section 3 though is dealing with research carried out entirely after WWII and could have been included in the fifth volume after its fifth section that covers similar matters. It may have been included in the fourth volume just to give it more substance.

2.2.5. *Tome V*

The fifth volume gathers together five sections published over the year 1948 to 1956 with P. Lacroute as Director of the Observatory. The printer was Abbadia Observatory in Hendaye.

- Fascicule 1 (1948):
  Mesures photométriques d'étoiles doubles, par P. Muller (pp. 3-19)
- Fascicule 2 (1949):
  Mesures photométriques d'étoiles doubles serrées, par P. Muller (pp. 3-23)
- Fascicule 3 (1951):
  Mesures photométriques d'étoiles doubles, par P. Muller (pp. 3-9)
  Magnitudes photométriques des composantes de 515 étoiles doubles ou multiples, par P. Muller (pp. 10-19)
- Fascicule 4 (1952):
  Colorimétrie de 51 étoiles doubles, par P. Muller (pp. 3-32)
- Fascicule 5 (1956):
  Table pour le calcul des éléments des orbites des étoiles doubles visuelles, par A. Danjon et P. Muller (pp. 3-28)
  Orbites des étoiles doubles ADS 3082, 6315, 9441, 11623, 13744, 15992, par P. Muller (pp. 29-36)
  Observations photoélectriques d'occultations, par P. Bacchus (pp. 37-43)
  Le nouvel équatorial de 210mm, par P. Lacroute (pp. 44-45)
  Le micromètre à double image de l'équatorial de 210mm, par P. Bru (p. 46)
  Observations d'étoiles doubles, par P. Bru et J. Huss (pp. 47-55)

2.2.6. *Tome VI*

This sixth and last volume of the French annals of Strasbourg Observatory is made of two sections published respectively in 1961 and 1964, under the directorship of P. Lacroute. The printer was again Abbadia Observatory in Hendaye.

- Fascicule 1 (1961):
  Le spectre de Nova Herculis 1960, par A. Florsch (pp. 5-16)
  Mesures d'étoiles doubles, par P. Bru (pp. 17-25)
- Fascicule 2 (1964):
  Étude des erreurs systématiques à craindre dans l'établissement d'un catalogue méridien relatif, par P. Lacroute (pp. 33-37)
  Amélioration dans les réductions d'un catalogue méridien. Méthode de "synthèse", par P. Lacroute (pp. 29-47)
  Réduction des observations méridiennes de Strasbourg avec la méthode de "synthèse", par P. Bacchus et P. Lacroute (pp. 49-74)
  La comparaison des catalogues, par P. Bacchus (pp. 75-83)
  Étude externe des observations méridiennes de Strasbourg, par P. Lacroute (pp. 85-95)
  Amélioration des constantes de clichés en utilisant leur recouvrement, par P. Lacroute (pp. 97-111)
  Étude sur l'établissement d'un catalogue photographique, par P. Lacroute (pp. 113-127)
  Les observations photographiques des satellites artificiels de la Terre à l'Observatoire de Strasbourg, par A. Schmitt (pp. 129-141)

3. Publications of Strasbourg Observatory

As it will be seen in the following, the generic title *Publications de l'Observatoire de Strasbourg* encompassed over the years quite a varying material: from reprints of papers in professional journals and communications at conferences to catalogs and full proceedings of in-house scientific meetings.

3.1. COLLECTIONS OF PAPERS

In spite of the fact that a number of the following papers have been published in professional journals, they are mentioned here for the record as they made up volumes of the series. Gathering together reprints in institutional collections has been fashionable for a while as this also contributed to exchanges. Librarians however quickly realized that this was increasingly duplicating subscriptions to journals (and eating up shelve space). The practice virtually disappeared in the early 1980s.

3.1.1. *Volume 1*

- Fascicule 1

  Lacroute, P. 1968, Études sur l'emploi de recouvrements de plaques pour l'établissement de catalogues photographiques, *Highlights in Astronomy*, D. Reidel Publ. Co., Dordrecht, 319-337.
- Fascicule 2

  Valbousquet, A. 1968, Mesures d'étoiles doubles à Strasbourg, *J. Observateurs* **51**, 21-26.
- Fascicule 3

  Lacroute, P. 1969, Étude des solutions optima pour des résolutions de clichés séparés, 87ème Congrès de l'Association Française pour l'Avancement des Sciences, Nancy, 11-23.
- Fascicule 4

  Schmitt, A. 1969, Méthode de réduction de la position d'un astre errant par rattachement à *n* étoiles de référence dont le centre de gravité est choisi voisin de l'astre, 87ème Congrès de l'Association Française pour l'Avancement des Sciences, Nancy, 33-38.
- Fascicule 5

  Schmitt, A. 1969, Discussion provisoire des résultats obtenus à l'Observatoire de Strasbourg dans l'observation des satellites géodésiques Echo I, Echo II et Pageos du Programme Géodésique Européen et Mondial, 87ème Congrès de l'Association Française pour l'Avancement des Sciences, Nancy, 28-32.
- Fascicule 6

  Schmitt, A. 1970, Précision des positions de satellites réduites à l'Observatoire de Strasbourg dans le Programme Géodésique Ouest Européen, Obs. Astron. Strasbourg, 1-7.

3.1.2. *Volume 2*

- Fascicule 1

  Florsch, A. 1972, Étude du petit nuage de Magellan par la technique du prisme-objectif de Fehrenbach, PhD Thesis, Obs. Astron. Strasbourg, 129 pp.
- Fascicule 2

  Schmitt, A. 1972, Formulae of Photographic Reduction and the Astronomical Refraction, Obs. Astron. Strasbourg, 1972, 2 pp.
- Fascicule 3

  Équipe Satellite et Géodésie 1972, Programme géodésique de l'Europe de l'Ouest. Station Strasbourg 05002. Tableau des erreurs moyennes quadratiques internes, Obs. Astron. Strasbourg, 2 pp.
- Fascicule 4

  Acker, A. 1971, Orbital Elements of the Spectroscopic Binaries HD

24733 and HD 861, *Astron. Astrophys.* **14**, 189-197.

– Fascicule 5

Schmitt, A. 1972, Principle of Satellite Position Determination by Look Angles from Two Stations Using the Great Circle of Simultaneousness and Determination of the Relative Systematic Error in Time Between the Two Stations, Obs. Astron. Strasbourg, 3 pp.

3.1.3. *Volume 3*

– Fascicule 1

Valbousquet, A. 1974, Étude de l'incertitude à craindre dans la résolution de catalogues photographiques utilisant le recouvrement de clichés, PhD Thesis, Obs. Astron. Strasbourg, 99 pp.

– Fascicule 2

Lacroute, P. & Valbousquet, A. 1974, Nouvelle résolution de l'AGK2-AGK3, *Astron. Astrophys. Suppl.* **16**, 445-510.

3.1.4. *Volume 4*

– Fascicule 1

Acker, A. 1976, Cinématique, âge et binarité des noyaux de nébuleuses planétaires, PhD Thesis, Obs. Astron. Strasbourg, 172 pp.

3.1.5. *Volume 5*

– Unpublished.

3.1.6. *Volume 6*

– Jaschek, M. & Jaschek, C. (Eds.) 1979, Réunion sur les étoiles Be, Strasbourg, 4 mai 1979, Obs. Astron. Strasbourg, 157 pp.

This volume is in fact made of the proceedings of the first *Journée de Strasbourg* (see next section).

3.2. JOURNÉES DE STRASBOURG

The proceedings of the *Journées de Strasbourg*, typically one-day meetings on specific themes, have made up a series on their own with the ISSN 0989-612X.

There have been 13 such gatherings organized yearly between 1979 and 1991. As mentioned in the previous section, the first proceedings were labeled as Volume 6 of the *Publications de l'Observatoire de Strasbourg* and have been repeated here for completeness of the following list.

1. Jaschek, M. & Jaschek, C. (Eds.) 1979, Réunion sur les étoiles Be (4 mai 1979), Obs. Astron. Strasbourg, 157 pp.

2. Jaschek, M. & Jaschek, C. (Eds.) 1980, Réunion sur les étoiles supergéantes (11 janvier 1980), Obs. Astron. Strasbourg, 185 pp.
3. Florsch, A., Jaschek, M. & Jaschek, C. (Eds.) 1981, Étoiles bleues en-dessous de la séquence principale (9 janvier 1981), Obs. Astron. Strasbourg, 134 pp.
4. Dubois, P., Florsch, A. & Jaschek, M. (Eds) 1982, La structure du Petit Nuage de Magellan (23 février 1982), Obs. Astron. Strasbourg, 92 pp.
5. Florsch, A., Jaschek, M. & Jaschek, C. (Eds.) 1983, Les étoiles binaires dans le diagramme HR (20 janvier 1983), Obs. Astron. Strasbourg, 103 pp.
6. Egret, D. & Guibert, J. (Eds.) 1984, L'avenir des données non stellaires (6 mars 1984), Obs. Astron. Strasbourg, 103 pp.
7. Florsch, A., Jaschek, M. & Jaschek, C. (Eds.) 1985, La composition chimique des étoiles dans le voisinage solaire (24 janvier 1985), Obs. Astron. Strasbourg, 98 pp.
8. Acker, A. (Ed.) 1986, Les nébuleuses planétaires (6 février 1986), Obs. Astron. Strasbourg, 111 pp. (ISBN 2-906361-00-3)
9. Fresneau, A. & Hamm, M. (Eds.) 1987, Impacts des surveys du visible sur notre connaissance de la galaxie (18 mars 1987), Obs. Astron. Strasbourg, 137 pp. (ISBN 2-906361-01-1)
10. Halbwachs, J.L., Jasniewicz, G. & Egret, D. (Eds.) 1988, Détection et classification des étoiles variables (21 avril 1988), Obs. Astron. Strasbourg, 99 pp. (ISBN 2-906361-02-X)
11. Heck, A. (Ed.) 1989, Artificial Intelligence Techniques for Astronomy – Techniques de l'intelligence artificielle pour l'astronomie (4 April 1989 – 4 april 1989), Obs. Astron. Strasbourg, 80 pp. (ISBN 2-906361-04-6)
12. Heck, A. (Ed.) 1990, Fractals in Astronomy (26 June 1990), *Vistas Astron.* **33** 247-424.
13. Jasniewicz, G. (Ed.) 1991, Stades avancés dans l'évolution des étoiles binaires serrées (23 mai 1991), Obs. Astron. Strasbourg, iii + 57 pp.

3.3. ASTRONOMIE ET SCIENCES HUMAINES

In 1986, P. Erny (Institute of Ethnology) and C. Jaschek initiated a series of meetings on astronomy and human sciences, the proceedings of which made up their own series with the ISSN 0989-6236. In total, 22 of such conferences took place, the last one in November 1996.

The following list is provided to the best of what could be extracted from archives and human memories. For instance, some conference dates remain incomplete (exact days missing). It should also be noted that, if the first volumes were entirely dedicated to specific meetings, the last ones gather together whatever could be collected from speakers at several gatherings. No communications were ever published for the 19th nor the 21st meetings.

1. Erny, P. & Jaschek, C. (Eds.) 1988, Inst. Ethnologie & Obs. Astron. Strasbourg, 125 pp. [Communications to the first three meetings held on 24 October 1986, 16 January 1987 and 22 May 1987.]
2. Erny, P. & Jaschek, C. (Eds.) 1988, Inst. Ethnologie & Obs. Astron. Strasbourg, 150 pp. [Communications to the 4th and 5th meetings held in December 1987 and April 1988.]
3. Erny, P. & Jaschek, C. (Eds.) 1989, Inst. Ethnologie & Obs. Astron. Strasbourg, 110 pp. [Communications to the 6th meeting held on 4 November 1988.]
4. Erny, P. & Jaschek, C. (Eds.) 1989, Inst. Ethnologie & Obs. Astron. Strasbourg, 147 pp. [Communications to the 7th meeting held on 28 April 1989.]
5. Erny, P. & Jaschek, C. (Eds.) 1990, Inst. Ethnologie & Obs. Astron. Strasbourg, 112 pp. [Communications to the 8th meeting.]
6. Erny, P. & Jaschek, C. (Eds.) 1991, Inst. Ethnologie & Obs. Astron. Strasbourg, 83 pp. [Communications to the 9th meeting.]
7. Erny, P. & Jaschek, C. (Eds.) 1992, Inst. Ethnologie & Obs. Astron. Strasbourg, 106 pp. [Communications to the 10th and 11th meetings.]
8. Erny, P. & Jaschek, C. (Eds.) 1992, Inst. Ethnologie & Obs. Astron. Strasbourg, 162 pp. [Communications to the 12th and 13th meetings.]
9. Jasniewicz, G. (Ed.) 1993, Inst. Ethnologie & Obs. Astron. Strasbourg, 168 pp. [Communications to the 14th and 15th meetings held on 6 November 1992 and 14 May 1993.]
10. Jasniewicz, G. (Ed.) 1994, Inst. Ethnologie & Obs. Astron. Strasbourg, 140 pp. [Communications to the 15th, 16th and 17th meetings held on 14 May 1993, 19 November 1993 and 27 May 1994.]
11. Jasniewicz, G. (Ed.) 1995, Inst. Ethnologie & Obs. Astron. Strasbourg, 88 pp. [Communications to the 17th and 18th meetings held on 27 May 1994 and 25 November 1994.]
12. Jasniewicz, G. (Ed.) 1996, Inst. Ethnologie & Obs. Astron. Strasbourg, 82 pp. [Communications to the 18th and 20th meetings held on 25 November 1994 and 24 November 1995.]
13. Valls-Gabaud, D. (Ed.) 1996, Inst. Ethnologie & Obs. Astron. Strasbourg, 136 pp. [Communications to the 20th and 22nd meetings held on 24 November 1995 and 26 November 1996.]

3.4. ÉQUIPE "POPULATION STELLAIRE"

The following publication, produced by one of the research teams (Stellar Populations), has been labeled separately:

– Bischoff, M., Florsch, A., Florsch, J. & Marcout, J. 1988, Catalogue des petits objets dans la direction du Petit Nuage de Magellan, Obs. Astron. Strasbourg, 250 pp. (ISBN 2-906361-03-8)

4. The CDS Information Bulletin

The Centre de Données Stellaires (CDS)[6] was installed at Strasbourg Observatory in 1972 (see Heck 2005b and Jung 2005 in this volume). In 1971, *i.e.* even before the official existence of the center, an *Information Bulletin* started to appear twice a year under the editorship of the CDS Director. The publication was discontinued after Issue 48 (March 1996).

The Bulletin carried two different titles over the years:
– from Issue 1 (January 1971) to Issue 12 (January 1977), it was known as *Centre de Données Stellaires – Information Bulletin*, while
– from Issue 13 (July 1977) onwards, that title switched to *Bulletin d'information du CDS* with the ISSN 0242-6536. From Issue 40 (January 1992) onwards, that ISSN became 1169-8837 as the new CDS explanation (see Footnote 6 – already in practical usage by the mid-1980s) appeared on the cover.

5. The CDS Special Publications

This section gathers together the various volumes published in the series entitled *Publications Spéciales du CDS* (CDS Special Publications) under the generic ISSN 0764-9614.

1. Fernandez, A., Le Squeren, A.M. & Lortet, M.C. 1977, Recherches bibliographiques sur la détection de raies moléculaires dans les objets galactiques – Bibliography of Molecular Lines in Galactic Objects, 161 pp. (ISBN 2-908064-00-6)
2. Fernandez, A. & Lortet, M.C. 1980, Detection and Measurements of Infrared Emission from Galactic Objets – Recherches bibliographiques sur la détection et les mesures de divers types d'objets galactiques en infrarouge, 136 pp. (ISBN 2-908064-01-4)
3. Acker, A., Chopinet, M., Gleizes, F., Marcout, J., Ochsenbein, F. & Roques, J.M. 1982, Catalog of the Central Stars of Planetary Nebulae (with the collaboration of Haute Provence Observatory), iv + 129 pp.
4. Jaschek, M. & Egret, D. 1982, Catalog of Stellar Groups: The Earlier Groups, 114 pp. (ISBN 2-908064-03-0)
5. Jaschek, M. & Egret, D. 1982, Catalog of Stellar Groups: The Later Groups (unpublished?)

[6]Later renamed as *Centre de Données astronomiques de Strasbourg.*

Figure 1. Stamps used by the library of Strasbourg Observatory over the years: a) first German stamp used in the 19th century until 1902-1903; b) second German stamp used both in the 19th and 20th centuries, sometimes together with the first stamp; c) third German stamp, used in the 20th century after 1906-1907; d) French stamp used after 1918; e) current stamp.

6. Bischoff, M. 1983, Tables de correspondances relatives au *Catalogue of Stellar Identifications*: 1ère partie, 77 pp. (ISBN 2-908064-04-9)
7. Bischoff, M. 1984, Tables de correspondances relatives au *Catalogue of Stellar Identifications*: 2ème partie, 73 pp. (ISBN 2-908064-05-7)
8. Heck, A. & Manfroid, J. 1985, International Directory of Astronomical Associations and Societies 1986 (IDAAS 1986), iv + 266 pp. (ISBN 2-908064-06-5)
9. Heck, A. & Manfroid, J. 1986, International Directory of Professional Astronomical Institutions 1987 (IDPAI 1987), iv + 276 pp. (ISBN 2-908064-07-3)
10. Heck, A. & Manfroid, J. 1987, International Directory of Astronomical Associations and Societies 1988 (IDAAS 1988), vi + 516 pp. (ISBN 2-908064-08-1)
11. Heck, A. 1988, A Bibliographical Catalogue of RR Lyrae Stars, 190 pp. (ISBN 2-908064-09-x)
12. Heck, A. 1988, International Directory of Professional Astronomical

Institutions 1989 (IDPAI 1989), vi + 492 pp. (ISBN 2-908064-10-3)
13. Heck, A. 1989, International Directory of Astronomical Associations and Societies together with related items of interest – Répertoire international d'associations et sociétés astronomiques ainsi que d'autres entrées d'intérêt général (IDAAS 1990), vi + 716 pp. (ISBN 2-908064-11-1)
14. Heck, A. 1989, International Directory of Professional Astronomical Institutions together with related items of interest – Répertoire international des institutions astronomiques professionnelles ainsi que d'autres entrées d'intérêt général (IDPAI 1990), vi + 658 pp. (ISBN 2-908064-12-x)
15. Heck, A. 1990, Acronyms and Abreviations in Astronomy and Space Sciences, ii + 254 pp. (ISBN 2-908064-13-8)
16. Heck, A. 1991, Astronomy, Space Sciences and Related Organizations of the World (ASpScROW 1991), x + 1182 pp. (two volumes) (ISBN 2-908064-14-6)
17. Marcout, J. & Florsch, A. 1990, Catalogue des objets dans la direction du Petit Nuage de Magellan, Complément n° 1 – Catalogue of Objects in the Direction of the Small Magellanic Cloud, 52 pp. (ISBN 2-908064-15-4)
18. Heck, A. 1991, Acronyms and Abreviations in Astronomy, Space Sciences and Related Fields, iv + 592 pp. (ISBN 2-908064-16-2)
19. Heck, A. 1992, StarBriefs 1992 – A Dictionary of Abbreviations, Acronyms, and Symbols in Astronomy, Space Sciences and Related Fields, xxxiv + 774 pp. (ISBN 2-908064-17-0)
20. Heck, A. 1993, StarGuides 1993 – A Directory of Astronomy, Space Sciences and Related Organizations of the World, x + 1174 pp. (two volumes) (ISBN 2-908064-14-6)
21. Heck, A. 1993, StarBriefs 1993 – A Dictionary of Abbreviations, Acronyms, and Symbols in Astronomy, Space Sciences and Related Fields, xxxiv + 778 pp. (ISBN 2-908064-19-7)
22. Heck, A. 1994, StarBriefs 1994 – A Dictionary of Abbreviations, Acronyms, and Symbols in Astronomy, Space Sciences and Related Fields, vi + 818 pp. (ISBN 2-908064-20-0)
23. Heck, A. 1994, StarGuides 1994 – A Directory of Astronomy, Space Sciences and Related Organizations of the World, viii + 880 pp. (ISBN 2-908064-21-9)
24. Lortet, M.C., Bordes, S. & Ochsenbein, F. 1994, The Second Reference Dictionnary of the Nomenclature of Celestial Objects, iii + 235 pp. (Volume I) and xii + 300 pp. (Volume II) (ISBN 2-908064-22-7)
25. Heck, A. 1995, StarGuides 1995 – A Directory of Astronomy, Space Sciences and Related Organizations of the World, viii + 814 pp. (ISBN

2-908064-23-5)

26. Heck, A. 1995, StarBriefs 1995 – A Dictionary of Abbreviations, Acronyms, and Symbols in Astronomy, Space Sciences and Related Fields, vi + 744 pp. (ISBN 2-908064-24-3)
27. Heck, A. 1996, StarGuides 1996 – A Directory of Astronomy, Space Sciences and Related Organizations of the World, viii + 916 pp. (ISBN 2-908064-25-1)
28. Heck, A. 1996, StarBriefs 1996 – A Dictionary of Abbreviations, Acronyms, and Symbols in Astronomy, Space Sciences and Related Fields, vi + 906 pp. (ISBN 2-908064-26-x)
29. Heck, A. 1997, StarGuides 1997 – A Directory of Astronomy, Space Sciences and Related Organizations of the World, viii + 1018 pp. (ISBN 2-908064-27-8)
30. Heck, A. 1998, StarGuides 1998 – A Directory of Astronomy, Space Sciences and Related Organizations of the World, viii + 1022 pp. (ISBN 2-908064-28-6)
31. Heck, A. 1999, StarGuides 1999 – A Directory of Astronomy, Space Sciences and Related Organizations of the World, viii + 1124 pp. (ISBN 2-908064-29-4)
32. Heck, A. 2000, StarGuides 2000 – A Directory of Astronomy, Space Sciences and Related Organizations of the World, viii + 1140 pp. (ISBN 2-908064-30-8)
33. Heck, A. 2000, StarBriefs 2000 – A Dictionary of Abbreviations, Acronyms, and Symbols in Astronomy, Space Sciences and Related Fields, vi + 802 pp. (ISBN 2-908064-31-6)

6. Newsletters

Some specialized newsletters were temporarily published at Strasbourg Observatory. They cannot be considered exactly as institutional publications, but are nevertheless mentioned here for the record.

6.1. BE STARS

The *Bulletin d'information sur les étoiles Be – The Be Star Newsletter* was edited twice yearly by M. Jaschek from 1980 to 1986 (Issues 1-13) for the Commission 29 of the International Astronomical Union (IAU). It carried the ISSN 0296-3140.

6.2. CHEMICALLY PECULIAR LATE-TYPE STARS

The *Bulletin sur les étoiles tardives à spectre particulier – The Newsletter of Chemically Peculiar Late-Type Stars* appeared twice yearly under the editorship of C. Jaschek from 1984 to 1987 (Issues 1 to 5) with the ISSN 0296-3132.

6.3. RADIAL VELOCITIES

The *Radial Velocities Newsletter* was also edited by C. Jaschek for IAU Commission 30, apparently only twice, in 1987 and 1988.

6.4. IAU WG ON ASTRONOMICAL DATA

The *Newsletter of the IAU Commision 5 Working Group on 'Astronomical Data'* was edited by A. Heck as Chairman of the WG from 1991 to 1994. Four issues were published.

6.5. WG FOR MODERN ASTRONOMICAL METHODOLOGY

The *Newsletter of the Working Group for Modern Astronomical Methodology* was edited by A. Heck and F. Murtagh (ST-ECF) between 1985 and 1989. Eight issues were published and were subsequently reprinted in the CDS Information Bulletin.

Acknowledgements

We thank H.W. Duerbeck for useful comments. Refer also to his chapter in this volume (Duerbeck 2005a) for *i.a.* a list of theses completed at Strasbourg Observatory during German times. Compiling a similar list for the French periods would have been very difficult: lack of appropriate data, frequent changes of the rules for theses in the French system (sometimes two different types of theses), plus not unfrequently theses prepared in Strasbourg, but presented elsewhere, and *vice versa*.

References

1. Duerbeck, W.H. 2005a, The Observatory of the Emperor Wilhelm University: The People behind the Documents, this volume.
2. Duerbeck, H. 2005b, Walter F. Wislicenus and Modern Astronomical Bibliography, this volume.
3. Esclangon, E. 1926, La Nouvelle Organisation de l'Observatoire, *Ann. Obs. Strasbourg* **I**, 1-44.
4. Heck, A. 2005a, The Multinational History of Strasbourg Observatory, this volume.
5. Heck, A. 2005b, Vistas into the CDS Genesis, this volume.
6. Jung, J. 2005, The Stellar Data Center: Origins and Early Beginnings (1972-1974), this volume.

Astrophysics and Space Science Library

Volume 324: ***Kristian Birkeland – The First Space Scientist,*** by A. Egeland, W.J. Burke. Hardbound ISBN 1-4020-3293-5, April 2005

Volume 323: ***Recollections of Tucson Operations,*** by M.A. Gordon. Hardbound ISBN 1-4020-3235-8, December 2004

Volume 322: ***Light Pollution Handbook,*** by K. Narisada, D. Schreuder Hardbound ISBN 1-4020-2665-X, November 2004

Volume 321: ***Nonequilibrium Phenomena in Plasmas***, edited by A.S. Shrama, P.K. Kaw. Hardbound ISBN 1-4020-3108-4, December 2004

Volume 320: ***Solar Magnetic Phenomena,*** edited by A. Hanslmeier, A. Veronig, M. Messerotti. Hardbound ISBN 1-4020-2961-6, December 2004

Volume 319: ***Penetrating Bars through Masks of Cosmic Dust***, edited by D.L. Block, I. Puerari, K.C. Freeman, R. Groess, E.K. Block. Hardbound ISBN 1-4020-2861-X, December 2004

Volume 318: ***Transfer of Polarized light in Planetary Atmospheres***, by J.W. Hovenier, J.W. Domke, C. van der Mee. Hardbound ISBN 1-4020-2855-5. Softcover ISBN 1-4020-2889-X, November 2004

Volume 317: ***The Sun and the Heliosphere as an Integrated System,*** edited by G. Poletto, S.T. Suess. Hardbound ISBN 1-4020-2830-X, November 2004

Volume 316: ***Civic Astronomy - Albany's Dudley Observatory, 1852-2002,*** by G. Wise
Hardbound ISBN 1-4020-2677-3, October 2004

Volume 315: ***How does the Galaxy Work - A Galactic Tertulia with Don Cox and Ron Reynolds,*** edited by E. J. Alfaro, E. Pérez, J. Franco
Hardbound ISBN 1-4020-2619-6, September 2004

Volume 314: ***Solar and Space Weather Radiophysics- Current Status and Future Developments,*** edited by D.E. Gary and C.U. Keller
Hardbound ISBN 1-4020-2813-X, August 2004

Volume 313: ***Adventures in Order and Chaos***, by G. Contopoulos. Hardbound ISBN 1-4020-3039-8, January 2005

Volume 312: ***High-Velocity Clouds,*** edited by H. van Woerden, U. Schwarz, B. Wakker
Hardbound ISBN 1-4020-2813-X, September 2004

Volume 311: ***The New ROSETTA Targets- Observations, Simulations and Instrument Performances***, edited by L. Colangeli, E. Mazzotta Epifani, P. Palumbo
Hardbound ISBN 1-4020-2572-6, September 2004

Volume 310: ***Organizations and Strategies in Astronomy 5,*** edited by A. Heck
Hardbound ISBN 1-4020-2570-X, September 2004

Volume 309: ***Soft X-ray Emission from Clusters of Galaxies and Related Phenomena***, edited by R. Lieu and J. Mittaz
Hardbound ISBN 1-4020-2563-7, September 2004

Volume 308: ***Supermassive Black Holes in the Distant Universe,*** edited by A.J. Barger
Hardbound ISBN 1-4020-2470-3, August 2004

Volume 307: ***Polarization in Spectral Lines***, by E. Landi Degl'Innocenti and M. Landolfi
Hardbound ISBN 1-4020-2414-2, August 2004

Volume 306: ***Polytropes – Applications in Astrophysics and Related Fields,*** by G.P. Horedt
Hardbound ISBN 1-4020-2350-2, September 2004

Volume 305: ***Astrobiology: Future Perspectives,*** edited by P. Ehrenfreund, W.M. Irvine, T. Owen, L. Becker, J. Blank, J.R. Brucato, L. Colangeli, S. Derenne, A. Dutrey, D. Despois, A. Lazcano, F. Robert
Hardbound ISBN 1-4020-2304-9, July 2004
Paperback ISBN 1-4020-2587-4, July 2004

Volume 304: ***Cosmic Gammy-ray Sources,*** edited by K.S. Cheng and G.E. Romero
Hardbound ISBN 1-4020-2255-7, September 2004

Volume 303: ***Cosmic rays in the Earth's Atmosphere and Underground***, by L.I, Dorman
Hardbound ISBN 1-4020-2071-6, August 2004

Volume 302:***Stellar Collapse,*** edited by Chris L. Fryer
Hardbound, ISBN 1-4020-1992-0, April 2004

Volume 301: ***Multiwavelength Cosmology***, edited by Manolis Plionis
Hardbound, ISBN 1-4020-1971-8, March 2004

Volume 300:***Scientific Detectors for Astronomy,*** edited by Paola Amico, James W. Beletic, Jenna E. Beletic
Hardbound, ISBN 1-4020-1788-X, February 2004

Volume 299: ***Open Issues in Local Star Fomation,*** edited by Jacques Lépine, Jane Gregorio-Hetem
Hardbound, ISBN 1-4020-1755-3, December 2003

Volume 298: ***Stellar Astrophysics - A Tribute to Helmut A. Abt,*** edited by K.S. Cheng, Kam Ching Leung, T.P. Li
Hardbound, ISBN 1-4020-1683-2, November 2003

Volume 297: ***Radiation Hazard in Space,*** by Leonty I. Miroshnichenko
Hardbound, ISBN 1-4020-1538-0, September 2003

Volume 296: ***Organizations and Strategies in Astronomy, volume 4,*** edited by André Heck
Hardbound, ISBN 1-4020-1526-7, October 2003

Volume 295: ***Integrable Problems of Celestial Mechanics in Spaces of Constant Curvature,*** by T.G. Vozmischeva
Hardbound, ISBN 1-4020-1521-6, October 2003

Volume 294: ***An Introduction to Plasma Astrophysics and Magnetohydrodynamics,*** by Marcel Goossens
Hardbound, ISBN 1-4020-1429-5, August 2003
Paperback, ISBN 1-4020-1433-3, August 2003

Volume 293: ***Physics of the Solar System,*** by Bruno Bertotti, Paolo Farinella, David Vokrouhlický
Hardbound, ISBN 1-4020-1428-7, August 2003
Paperback, ISBN 1-4020-1509-7, August 2003

Volume 292: ***Whatever Shines Should Be Observed,*** by Susan M.P. McKenna-Lawlor
Hardbound, ISBN 1-4020-1424-4, September 2003

Volume 291: ***Dynamical Systems and Cosmology***, by Alan Coley
Hardbound, ISBN 1-4020-1403-1, November 2003

Volume 290: ***Astronomy Communication,*** edited by André Heck, Claus Madsen
Hardbound, ISBN 1-4020-1345-0, July 2003

Volume 287/8/9: ***The Future of Small Telescopes in the New Millennium,*** edited by Terry D. Oswalt
Hardbound Set only of 3 volumes, ISBN 1-4020-0951-8, July 2003

Volume 286: ***Searching the Heavens and the Earth: The History of Jesuit Observatories,*** by Agustín Udías
Hardbound, ISBN 1-4020-1189-X, October 2003

Volume 285: ***Information Handling in Astronomy - Historical Vistas,*** edited by André Heck
Hardbound, ISBN 1-4020-1178-4, March 2003

Volume 284: ***Light Pollution: The Global View,*** edited by Hugo E. Schwarz
Hardbound, ISBN 1-4020-1174-1, April 2003

Volume 283: ***Mass-Losing Pulsating Stars and Their Circumstellar Matter,*** edited by Y. Nakada, M. Honma, M. Seki
Hardbound, ISBN 1-4020-1162-8, March 2003

Volume 282: ***Radio Recombination Lines,*** by M.A. Gordon, R.L. Sorochenko
Hardbound, ISBN 1-4020-1016-8, November 2002

Volume 281: ***The IGM/Galaxy Connection,*** edited by Jessica L. Rosenberg, Mary E. Putman
Hardbound, ISBN 1-4020-1289-6, April 2003

Volume 280: ***Organizations and Strategies in Astronomy III,*** edited by André Heck
Hardbound, ISBN 1-4020-0812-0, September 2002

Volume 279: ***Plasma Astrophysics , Second Edition,*** by Arnold O. Benz
Hardbound, ISBN 1-4020-0695-0, July 2002

Volume 278: ***Exploring the Secrets of the Aurora,*** by Syun-Ichi Akasofu
Hardbound, ISBN 1-4020-0685-3, August 2002

Volume 277: ***The Sun and Space Weather,*** by Arnold Hanslmeier
Hardbound, ISBN 1-4020-0684-5, July 2002

Volume 276: ***Modern Theoretical and Observational Cosmology,*** edited by Manolis Plionis, Spiros Cotsakis
Hardbound, ISBN 1-4020-0808-2, September 2002

Volume 275: ***History of Oriental Astronomy,*** edited by S.M. Razaullah Ansari
Hardbound, ISBN 1-4020-0657-8, December 2002

Volume 274: ***New Quests in Stellar Astrophysics: The Link Between Stars and Cosmology,*** edited by Miguel Chávez, Alessandro Bressan, Alberto Buzzoni,Divakara Mayya
Hardbound, ISBN 1-4020-0644-6, June 2002

Volume 273: ***Lunar Gravimetry,*** by Rune Floberghagen
Hardbound, ISBN 1-4020-0544-X, May 2002

Volume 272:***Merging Processes in Galaxy Clusters,*** edited by L. Feretti, I.M. Gioia, G. Giovannini
Hardbound, ISBN 1-4020-0531-8, May 2002

Volume 271: ***Astronomy-inspired Atomic and Molecular Physics,*** by A.R.P. Rau
Hardbound, ISBN 1-4020-0467-2, March 2002

Volume 270: ***Dayside and Polar Cap Aurora,*** by Per Even Sandholt, Herbert C. Carlson, Alv Egeland
Hardbound, ISBN 1-4020-0447-8, July 2002

Volume 269: ***Mechanics of Turbulence of Multicomponent Gases,*** by Mikhail Ya. Marov, Aleksander V. Kolesnichenko
Hardbound, ISBN 1-4020-0103-7, December 2001

Volume 268: ***Multielement System Design in Astronomy and Radio Science,*** by Lazarus E. Kopilovich, Leonid G. Sodin
Hardbound, ISBN 1-4020-0069-3, November 2001

Volume 267: ***The Nature of Unidentified Galactic High-Energy Gamma-Ray Sources,*** edited by Alberto Carramiñana, Olaf Reimer, David J. Thompson
Hardbound, ISBN 1-4020-0010-3, October 2001

Volume 266: ***Organizations and Strategies in Astronomy II,*** edited by André Heck
Hardbound, ISBN 0-7923-7172-0, October 2001

Volume 265: ***Post-AGB Objects as a Phase of Stellar Evolution,*** edited by R. Szczerba, S.K. Górny
Hardbound, ISBN 0-7923-7145-3, July 2001

Volume 264: ***The Influence of Binaries on Stellar Population Studies,*** edited by Dany Vanbeveren
Hardbound, ISBN 0-7923-7104-6, July 2001

Volume 262: ***Whistler Phenomena - Short Impulse Propagation,*** by Csaba Ferencz, Orsolya E. Ferencz, Dániel Hamar, János Lichtenberger
Hardbound, ISBN 0-7923-6995-5, June 2001

Volume 261: ***Collisional Processes in the Solar System,*** edited by Mikhail Ya. Marov, Hans Rickman
Hardbound, ISBN 0-7923-6946-7, May 2001

Volume 260: ***Solar Cosmic Rays,*** by Leonty I. Miroshnichenko
Hardbound, ISBN 0-7923-6928-9, May 2001

Volume 259: ***The Dynamic Sun,*** edited by Arnold Hanslmeier, Mauro Messerotti, Astrid Veronig
Hardbound, ISBN 0-7923-6915-7, May 2001

Volume 258: ***Electrohydrodynamics in Dusty and Dirty Plasmas- Gravito-Electrodynamics and EHD,*** by Hiroshi Kikuchi
Hardbound, ISBN 0-7923-6822-3, June 2001

Volume 257: ***Stellar Pulsation - Nonlinear Studies,*** edited by Mine Takeuti, Dimitar D. Sasselov
Hardbound, ISBN 0-7923-6818-5, March 2001

Volume 256: ***Organizations and Strategies in Astronomy,*** edited by André Heck
Hardbound, ISBN 0-7923-6671-9, November 2000

Volume 255: ***The Evolution of the Milky Way- Stars versus Clusters,*** edited by Francesca Matteucci, Franco Giovannelli
Hardbound, ISBN 0-7923-6679-4, January 2001

Volume 254: ***Stellar Astrophysics***, edited by K.S. Cheng, Hoi Fung Chau, Kwing Lam Chan, Kam Ching Leung
Hardbound, ISBN 0-7923-6659-X, November 2000

Volume 253: ***The Chemical Evolution of the Galaxy***, by Francesca Matteucci
Paperback, ISBN 1-4020-1652-2, October 2003
Hardbound, ISBN 0-7923-6552-6, June 2001

Volume 252: ***Optical Detectors for Astronomy II***, edited by Paola Amico, James W. Beletic
Hardbound, ISBN 0-7923-6536-4, December 2000

Volume 251: ***Cosmic Plasma Physics***, by Boris V. Somov
Hardbound, ISBN 0-7923-6512-7, September 2000

Volume 250: ***Information Handling in Astronomy***, edited by André Heck
Hardbound, ISBN 0-7923-6494-5, October 2000

Volume 249: ***The Neutral Upper Atmosphere***, by S.N. Ghosh
Hardbound, ISBN 0-7923-6434-1, July 2002

Volume 247: ***Large Scale Structure Formation***, edited by Reza Mansouri, Robert Brandenberger
Hardbound, ISBN 0-7923-6411-2, August 2000

Volume 246: ***The Legacy of J.C. Kapteyn***, edited by Piet C. van der Kruit, Klaas van Berkel
Paperback, ISBN 1-4020-0374-9, November 2001
Hardbound, ISBN 0-7923-6393-0, August 2000

Volume 245: ***Waves in Dusty Space Plasmas***, by Frank Verheest
Paperback, ISBN 1-4020-0373-0, November 2001
Hardbound, ISBN 0-7923-6232-2, April 2000

Volume 244: ***The Universe***, edited by Naresh Dadhich, Ajit Kembhavi
Hardbound, ISBN 0-7923-6210-1, August 2000

Volume 243: ***Solar Polarization***, edited by K.N. Nagendra, Jan Olof Stenflo
Hardbound, ISBN 0-7923-5814-7, July 1999

Volume 242: ***Cosmic Perspectives in Space Physics***, by Sukumar Biswas
Hardbound, ISBN 0-7923-5813-9, June 2000

Volume 241: ***Millimeter-Wave Astronomy: Molecular Chemistry & Physics in Space,*** edited by W.F. Wall, Alberto Carramiñana, Luis Carrasco, P.F. Goldsmith
Hardbound, ISBN 0-7923-5581-4, May 1999

Volume 240: ***Numerical Astrophysics,*** edited by Shoken M. Miyama, Kohji Tomisaka,Tomoyuki Hanawa
Hardbound, ISBN 0-7923-5566-0, March 1999

Volume 239: ***Motions in the Solar Atmosphere,*** edited by Arnold Hanslmeier, Mauro Messerotti
Hardbound, ISBN 0-7923-5507-5, February 1999

Volume 238: ***Substorms-4,*** edited by S. Kokubun, Y. Kamide
Hardbound, ISBN 0-7923-5465-6, March 1999

For further information about this book series we refer you to the following web site:
www.springeronline.com

To contact the Publishing Editor for new book proposals:
Dr. Harry (J.J.) Blom: harry.blom@springer-sbm.com

FSC
www.fsc.org
MIX
Papier | Fördert gute Waldnutzung
FSC® C083411

Kolibri 360 GmbH
Ferdinand-Jühlke-Straße 7
99095 Erfurt, Deutschland
produktsicherheit@kolibri360.de